Green Energy and Technology

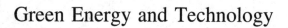

For further volumes:
http://www.springer.com/series/8059

A. B. M. Shawkat Ali
Editor

Smart Grids

Opportunities, Developments, and Trends

 Springer

Editor
A. B. M. Shawkat Ali
School of Information and Communication Technology
Central Queensland University
North Rockhampton, QLD
Australia

ISSN 1865-3529   ISSN 1865-3537 (electronic)
ISBN 978-1-4471-6186-8   ISBN 978-1-4471-5210-1 (eBook)
DOI 10.1007/978-1-4471-5210-1
Springer London Heidelberg New York Dordrecht

Printed on acid-free paper

Springer is part of Springer Science+Business Media (www.springer.com)

Dedicated to my lovely mother
Mrs. Soufia Khatun

Contents

1   **The Traditional Power Generation and Transmission System:
Some Fundamentals to Overcome Challenges**   1
Md Fakhrul Islam, Amanullah M. T. Oo
and Shaheen Hasan Chowdhury

2   **Smart Grid**   23
Md Rahat Hossain, Amanullah M. T. Oo and A. B. M. Shawkat Ali

3   **Renewable Energy Integration: Opportunities and Challenges** ...   45
G. M. Shafiullah, Amanullah M. T. Oo, A. B. M. Shawkat Ali,
Peter Wolfs and Mohammad T. Arif

4   **Energy Storage: Applications and Advantages**   77
Mohammad Taufiqul Arif, Amanullah M. T. Oo
and A. B. M. Shawkat Ali

5   **Smart Meter**   109
M. Rahman and Amanullah M. T. Oo

6   **Demand Forecasting in Smart Grid**   135
A. B. M. Shawkat Ali and Salahuddin Azad

7   **Database Systems for the Smart Grid**   151
Zeyar Aung

8   **Securing the Smart Grid: A Machine Learning Approach**   169
A. B. M. Shawkat Ali, Salahuddin Azad and Tanzim Khorshed

**9   Smart Grid Communication and Networking Technologies:
      Recent Developments and Future Challenges**............... 199
      Faisal Tariq and Laurence S. Dooley

10   Economy of Smart Grid............................. 215
      Gang Liu, M. G. Rasul, M. T. O. Amanullah and M. M. K. Khan

Index 229

Chapter 1
The Traditional Power Generation and Transmission System: Some Fundamentals to Overcome Challenges

Md Fakhrul Islam, Amanullah M. T. Oo
and Shaheen Hasan Chowdhury

Abstract In present power system, the engineers face variety of challenges in planning, construction and operation. In some of the problems, the engineers need to use managerial talents. In system design or upgrading the entire system into automatic control instead of slow response of human operator, the engineers need to exercise more technical knowledge and experience. It is principally the engineer's ability to achieve the success in all respect and provide the reliable and uninterrupted service to the customers. This chapter covers some important areas of the traditional power system that helps engineers to overcome the challenges. It emphasizes the characteristics of the various components of a power system such as generation, transmission, distribution, protection and SCADA system. During normal operating conditions and disturbances, the acquired knowledge will provide the engineers the ability to analyse the performance of the complex system and execute future improvement.

1.1 Introduction

Electric power system is a network that consists of electrical machines, lines and mechanism to generate electricity and supply to the customers. For a small region, this generation and supply of electricity are simple but it cannot ensure the

M. F. Islam (✉)
Deakin University, Geelong, Australia
e-mail: Fakhrul.Islam@ieee.org

A. M. T. Oo
School of Engineering, Deakin University, Geelong, Australia
e-mail: aman.m@deakin.edu.au

S. H. Chowdhury
Higher Education Division CQ University, Rockhampton, Australia

A. B. M. S. Ali (ed.), *Smart Grids*, Green Energy and Technology,
DOI: 10.1007/978-1-4471-5210-1_1, © Springer-Verlag London 2013

consistency of the supply and may cost much higher. In the modern world, the generators and small networks are interconnected and electrical energy can be transmitted to a long distance and finally supply to the industries, hospitals, commercial buildings and dwellings in different regions. This type of interconnected power network is known as grid. The grid is a very complex system, and the engineers face variety of challenges in planning, construction and operation of this system. The entire power system is divided into sections such as generation, transmission and distribution. This chapter discusses the important aspects in all the three sections of the traditional power system. Voltage, reactive power control and the protection system are important aspects and relate to the system stability and major interruptions. Supervisory control and data acquisition (SCADA) system has become an important area which plays critical role in day-to-day operations of power system. It also manages to perform the activities faster. This chapter emphasizes on these sections and illustrates to get good understanding for solving the future problems. It is expected that the subject matter in this chapter will upgrade the engineer's technical knowledge and enable to provide reliable and uninterrupted power supply to the consumers.

1.2 Power System

The interconnected power network is generally divided into three sectors such as Generation, Transmission and Distribution. Figure 1.1 represents a typical uncomplicated single line diagram of a power system. The sections of the power system are connected via transformers (Tx) which step up or step down the voltages.

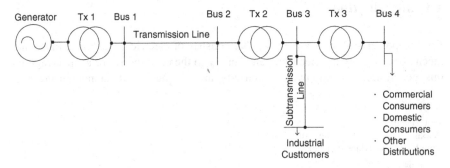

Fig. 1.1 Single line diagram of an uncomplicated power system with busbar (*Bus*) and transformer (Tx)

1.2.1 Generation

In generation section, the generators convert the energy from various sources to electric power at a particular voltage. Power generator has a prime mover and excitation system to control the output power as per the demand. For economic dispatch of power generation, control centre collects all the generator outputs and manages to control the output values based on the individual unit's generation costs. With system load variation, the generator outputs are controlled to have changes according to their individual generation cost to meet the system demand. Some instantly generating ability is always maintained by the individual generators for taking over sudden load for any unforeseen outage of generators or abrupt increase of the demand. Generation frequency is an indication of this sudden load changes. AC electric power is generated at frequencies 50 or 60 cycles per second (cs) which is also known as hertz (Hz). For example, Australia produces electric power at 50 Hz, and USA produces it at 60 Hz. This frequency remains constant at steady-state condition when the mechanical torque balances the electrical torque and the torque loss (T_{loss}) due to friction and windage. Mechanical torque (T_m) is produced by the prime mover for the applied force to its input from the energy source. Electrical torque (T_e) is generated as a result of the rotating electromagnetic field of the armature current. The torque equation at the generator output is given as

$$T_m - T_e = T_{loss}. \tag{1.1}$$

Since T_{loss} is very small in comparison with the other, it can be neglected. Hence, it can be stated that the generator operates at a constant speed in the steady state when the difference between the mechanical torque and electrical torque is zero. If sudden electrical load increases, the electrical torque also increases. At this stage, if the mechanical torque remains constant, the rotor mass decelerates causing a decrease in frequency, that is, the system frequency drops when $T_m < T_e$. Similarly, if the mechanical torque remains constant for sudden decrease in electrical load, the rotor mass accelerates resulting an increase in frequency, that is, the system frequency rises when $T_m > T_e$.

In order to maintain the constant system frequency, the mechanical torque needs to be increased or decreased with respect to the changes in electrical torque due to the electric load variation. This variation of mechanical torque is done by increasing or decreasing the input energy to the prime mover and is known as torque for acceleration (T_a). So, for negligible frictional loss, the torque equation of a generator can be written as in Eq. (1.2). As per the equation, T_a becomes zero when the generator attains its synchronous speed.

$$T_m - T_e = T_a. \tag{1.2}$$

Multiplying the angular speed (w) on both sides of the equation, the power balance equation is given as in Eq. (1.3).

$$P_m - P_e = P_a \qquad (1.3)$$

where $P_m = wT_m$ is the mechanical power supplied by the prime mover, $P_e = wT_m$ the electromagnetic power of the generator, and $P_a = wT_a$ the increase in power input to the prime mover for acceleration. The mechanism that controls the inputs of the prime mover according to generator load variation to preserve the synchronous speed and frequency is known as governor of the power generating system. There are varieties of prime movers and governors. Mostly used conventional prime movers are as follows;

- steam turbine prime mover,
- hydraulic turbine prime mover,
- combustion turbine prime mover,
- combined cycle prime mover.

The modern complex power generation systems are making use of them according to their efficiency and applicability.

In addition to the prime mover and governor, the generator has exciter and voltage regulator to control the output voltage at the generator terminals as well as the reactive power and real power flow. Exciters are basically DC (direct current) generators which supply DC current to the field winding usually on the rotor. There are also various AC (alternate current) excitation systems that use solid rectifiers to supply DC current to the field windings.

1.2.2 Transmission

Electric power is transmitted to a long distance through high-voltage (HV) overhead lines or HV underground cables. The range of transmission line voltages are 33 to 500 kV. Some power consumers specially the industries are provided power supply from the 33 to 66 kV transmission lines. These power consumers have their own substations to step down the voltage level according to their requirements. Power transmission at these 33 to 66 kV voltage levels is known as sub-transmission system.

The enormous electrical power is transmitted using 132 to 500 kV transmission lines. The voltage levels are determined on the basis of the amount of power and the distance it has to be transmitted. Power transmission network is highly meshed, though few radial transmission and/or sub-transmission lines in many rural and developing areas are found for supplying power to small inhabitants [1]. The reason of the meshed network is to ensure the non-interrupted power supply due to loss of any section of the network.

In the meshed network system designed for secured electric supply, the lines and transformers are underutilized than their capacities. They are generally loaded around 50 % to allow continuation of supply after outage of the load sharing lines or transformers. This underutilization of the transmission system increases the power transmission cost. The other factors of limiting the maximum transmission capacity of the system are voltage magnitude limits, the real power generation of the power plant and appropriate reactive power support. In order to get the benefit of meshed system, sufficient generation need to be ensured to meet peak demands. There should have a balance of demand and supply, so that the schedule power curtailments are not required. However, besides the supply reliability, the meshed network provides some more benefits such as broad choices of generating plants and their locations, reduction in reserve capacity of generators, diversity of load demand and many others.

The power electronics contributes to the development of high-voltage direct current (HVDC) transmission system [2–4]. At the same time, it contributes for the development of flexible AC transmission system (FACTS) as detail in [5, 6]. HVDC transmission system needs some additional equipment which increases the initial investments, but it offers lower transmission loss and cost compared to AC system when used for large amount of energy transfer at long distances. In selecting the HVDC system, it is important to compare the total capital and operating costs with the high-voltage AC system (HVAC) system.

High-voltage DC power transmission started in 1984 with 300 kV and in 1985 with 600 kV at the Itaipu hydropower plant in Brazil (Rudervall et al. [7]). In those times, HVDC is chosen over an AC system for transmission of power more than 500 MW at distances greater than 500 km. With the advancement of technology and materials, in today's electricity industry, HVDC transmission systems are being in use at very low voltages compared to old installations. For example, in Australia, 80 kV DC cables are used to connect the Queensland and New South Wales electricity grids between Terranora and Mullumbimby in the year 2000 for the transmission of 180 MVA load over a distance of 65 km. There are reasons of choosing HVDC transmission system over an AC system. Some of the important reasons are given below:

- technically efficient for long distance power transfer,
- offer low environmental impact and better power quality,
- allow interconnection of two networks in an asynchronous manner,
- imply stability improvements.

1.2.3 Distribution

Distribution is the power supply lines between high-voltage substations and customers. Mostly the distribution voltage level is 11 kV, but in many circumstances, voltages are higher up to 66 kV. They are step down to 440 V (line to line voltage) through distribution transformers before supplying to the customers. Some commercial customers receive three-phase 440 V supply. The other commercial customers and domestic customers receive single-phase 250 V supply. Conventional 440 V supply lines in the distribution are radial. Higher-voltage distribution lines are also rarely interconnected unless there are essential services. So, in distribution

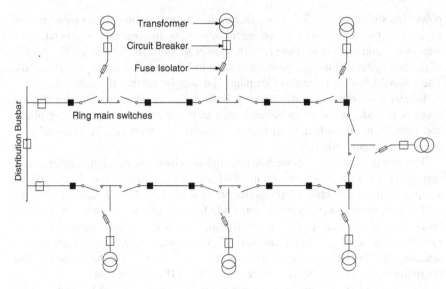

Fig. 1.2 Arrangement of ring-main feeder in distribution system

network in normal circumstances, a number of consumers may experience loss of supply for the failure of a section of the network.

In order to reduce the duration of the supply interruptions to the consumers, the auto-reclosers and sectionalizers are used in the radial lines/feeders. In business and densely populated areas, sometimes 11 kV supply network is arranged in ring-main feeders from the same switching station/substation with an open point preferably in the middle, so that for fault in one section, the consumers can be provided supply from other side. An example of ring-main feeders is shown in Fig. 1.2. This arrangement improves power restoration time, but the initial supply interruptions still take place. The reason of arranging the feeder rings from same substation is that the distribution feeders are not normally fitted with synchronizing equipment.

1.3 Power System Reliability and Quality

The reliability of power supply is the capability of delivering the uninterrupted power to consumers within accepted standards and in the amount. It relates more to complete loss of voltages rather than sags, swells, impulses or harmonics [8]. Reliability can be measured by frequency and duration of interruptions as well as magnitude of adverse effects on the electric supply.

Power quality may be defined as the measures, analysis and improvement of bus voltage to maintain that voltage to be sinusoidal at rated voltage and frequency, so that it is suitable for the operations of the consumers' devices [8, 9].

Modern sophisticated electronic equipments are sensitive to the events such as voltage sags, impulses, harmonics and phase imbalances. These events contaminate the sinusoidal wave of the voltage and hence become the power quality concern. Power quality events may arise from different causes such as switching or tripping of large inductive loads, lighting strikes and fault in the network. Some of the events such as harmonics may originate from customer's own installation due to the load temperament and impure the network.

Both power system reliability and power quality have a huge economic impact. There are huge production losses in consumer's industries. Process industry can be particularly vulnerable to problems with voltage sags and unbalance conditions because the equipments are interconnected with different controls, relays and adjustable speed drives where a trip of any component in the process can cause the whole plant to shut down [10]. The losses in process industries are also much higher because constituent materials may become irrecoverable scrap once the particular equipment or total industry trips during the production. Moreover, a process interruption for short duration (less than a minute) may cause a complete restart of the industry accruing to hours of production loss. Besides the total losses in industrial production, the equipment also experience damages and reduction in life expectancy [11].

Voltage sag may drop below the nominal voltage by 10 to 90 %, and its duration is 0.5 cycles to 1 min. Voltage drops below 90 % of nominal voltage for more than 1 min is known as under-voltage. Over-voltage occurs when nominal voltage rises above 110 % for more than 1 min. These over-voltage and under-voltage may occur and last for longer duration at the steady state of the system. These over-voltage and under-voltage at steady state are sometimes expressed as high steady-state voltage and low steady-state voltage. The major causes of steady-state voltage problems are overloading, inappropriate design, load switching and faulty regulating equipment. High steady-state voltages shrink the life of electronic devices and light bulbs. Alternatively, a low steady-state voltage reduces illumination levels, shirks the television pictures and causes slow heating of heating devices, motor starting problems and overheating in motors [12, 13].

1.4 Voltage Profile of Power System

1.4.1 Load Characteristics

Power system analysis involves relevant models of its components such as generation, transmission and distribution, and load devices. Knowledge of the load characteristics has a significant effect on loan modelling to analyse the load flow in the system. Consumer loads are categorized as follows.

Constant impedance load: Constant impedance load is a static load where the power varies with the square of the voltage magnitude. It can also be termed as constant admittance load.

Constant current load: Constant current load is a static load where the power varies directly with voltage magnitude.

Constant power load: Constant power load is a static load where the power does not vary with changes in voltage magnitude. It can also be termed as constant MVA load. Constant power load can be expressed as in Eq. (1.4).

$$P = P_0 \left(\frac{V}{V_0}\right)^x, \quad Q = Q_0 \left(\frac{V}{V_0}\right)^y \tag{1.4}$$

where P_0 and Q_0 are the real and reactive powers at a reference voltage V_0. Similarly, P and Q are the real and reactive power at a reference V. The exponents x and y depend on the type of load, for example, for constant power load $x = y = 0$, for constant current load $x = y = 1$ and for constant impedance load $x = y = 2$.

Polynomial load: Polynomial load is a static load that represents the power–voltage relationship as a polynomial equation of voltage magnitude. It is made up of constant impedance, constant current and constant power loads. A polynomial power load can be shown as in Eqs. (1.5) and (1.6).

$$P = P_0 \left[a_P \left(\frac{V}{V_0}\right)^2 + b_P \left(\frac{V}{V_0}\right) + c_P \right] \tag{1.5}$$

$$Q = Q_0 \left[a_Q \left(\frac{V}{V_0}\right)^2 + b_Q \left(\frac{V}{V_0}\right) + c_Q \right] \tag{1.6}$$

where a_P b_P, c_P, a_Q b_Q, c_Q are fractional values such that $(a_P + b_P + c_P) = (a_Q + b_Q + c_Q) = 1$.

1.4.2 Power Transfer Through Radial Feeder

Transmission or feeder lines transferring the power have a voltage drop between sending end (source bus) and the receiving end (load bus) voltages. Determination of this voltage drop for a radial feeder, supplying power from a source point, is an easy solution, and it depends on phase angle, voltage dependency of static load and line parameters. If the V_S and V_L are the voltages of source and load buses their vector relations with voltage drop V_d and the load current I_L can be presented as in Fig. 1.3.

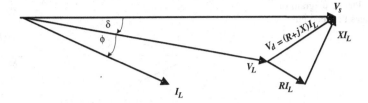

Fig. 1.3 Vector relations of the current and voltages in a radial feeder

where R and X are the equivalent resistance and reactance of the line, δ is the phase angle between V_S and V_L, and ϕ is the phase angle between voltage and current at load bus.

Consumer load phase angle ϕ may vary for the addition of load depending on its characteristic. For the increase of ϕ, the current will increase which reduces the V_L, and the decrease of ϕ will have opposite effects. In power system, Cos (ϕ) is known as the power factor (pf) of the load.

1.4.3 Power Transfer Between Active Sources

In a mesh transmission system, load flow is complicated. In that case, interconnecting buses can be the active source of synchronous system or generators. Power will flow from the bus which has the voltage that leads the voltage of the other bus. Figure 1.4 shows a simple diagram of the two synchronous buses where bus voltage V_1 leads the bus voltage V_2 by the power angle δ and V_2 presents the reference voltage. Considering the resistance of the transmission line is very low comparing to the reactance, it has been neglected to simplify the analysis. The phasor diagram of the voltages between bus 1 and bus 2 is presented in Fig. 1.5.

Voltage between bus 1 and bus 2 is

$$XI_2\sin\phi + jXI_2 \cos \phi = V_1\angle\delta - V_2\angle 0. \qquad (1.7)$$

Expressing all in phasor components

$$XI_2\sin\phi + jXI_2 \cos \phi = V_1 \cos \delta + jV_1 \sin \delta - V_2. \qquad (1.8)$$

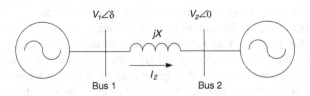

Fig. 1.4 Interconnection between two synchronous systems

Fig. 1.5 Phasor diagram of
the voltages between bus 1
and bus 2

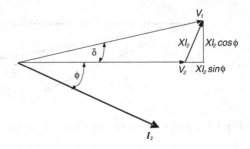

Multiplying both sides by $-jV_2/X$

$$V_2I_2 \cos \phi - jV_2I_2\sin\phi = \frac{V_2V_1 \sin \delta - jV_2(V_1 \cos \delta - V_2)}{X}. \qquad (1.9)$$

By equating the real and the reactive terms of Eq. (1.9), the real power P and reactive power Q can be expressed as

$$P = V_2I_2 \cos \phi = \frac{V_2V_1 \sin \delta}{X} \qquad (1.10)$$

$$Q = V_2I_2\sin\phi = \frac{V_2(V_1 \cos \delta - V_2)}{X}. \qquad (1.11)$$

1.5 Power System Stability and Control

The modern society needs ever-increasing electric power to meet the industrial commercial and domestic demands. In order to meet this demand, the generation, transmission and distribution capacities are increasing very fast forming a gigantic, complex interconnected power system. The main issues of this vast system are the voltage regulation, stability, protection system to isolate the section under fault and hence ensure continuity of the supply system.

Power networks always experience random changes in load due to faults on the network, switching in or out a large load such as a steel mill or heavy industry, loss of lines or generating units. These changes can be considered as the shifts of the power system from one stable state condition to another. During the time of shifting between two stable states, the system undergoes dynamics of the transition during which generator may lose synchronism or oscillate and finally trips. There is also a possibility of tripping line circuit breaker causing a power loss. So, during any disturbance, the power, frequency and voltage changes need to be subsequently adjusted by the generators during and after the transition. For example, in case a generator trips, the other connected generators must be capable of sharing the load demand; or if a line is lost, the power it was carrying must be supplied through the alternate sources. Failure of balancing the power may lead to a catastrophe of cascade tripping of generators and sections of the lines resulting disintegration of power network. Many control devices response to adjust the voltage, frequency and load sharing by generators during transition time of power system disturbances. Distance relay used for the line protection in the transmission system has the power swing module that can pick up during the power swing between the generators and block the line tripping. In case generators are overloaded resulting in slowing down the speed, frequency relays using for load shedding can isolate some unimportant feeders and save the generators from cascade tripping. Generator exciters, voltage regulating mechanism and governor play the vital role in stabilizing the transition time problems.

1.5.1 Generator Excitation

Excitation of a generator controls the electromagnetic force (EMF) of the generator which sequentially controls output voltage, power factor and current magnitude. Figure 1.6 shows a synchronous generator with simple excitation system.

An initial excitation induces a generator internal voltage E_g and delivers a power P at load bus voltage V and current I at power factor (pf) $\cos\phi$ across a generator reactance X. The power for this excitation can be expressed as in Eqs. (1.12) and (1.13). When the excitation is increased while the input power/torque is kept the same by the governor and voltage (V) is maintained equal by other machines operating in parallel, the generator internal voltage, current, phase angle and power angle will be changed to new values as expressed in Eqs. (1.14) and (1.15).

$$P = \frac{E_g V \sin \delta}{X} \tag{1.12}$$

$$P = VI \cos \phi \tag{1.13}$$

$$P = \frac{E_{g1} V \sin \delta_1}{X} \tag{1.14}$$

$$P = VI_1 \cos \phi_1. \tag{1.15}$$

The phasor diagram of the voltage, currents, phase and power angles for these two excitation conditions is shown in Fig. 1.7.

Fig. 1.6 Synchronous generators with simple excitation

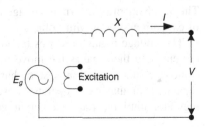

Fig. 1.7 Excitation for
increasing voltage E_g with
constant power and voltage

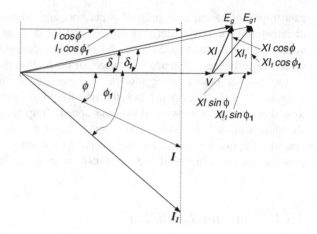

As observed in Fig. 1.7, due to the increase in excitation generator, internal voltage is increases to E_{g1}, and power angle is decreased to δ_1 to maintain the constant power. Similarly, current is increase to I_1 and the phase angle is increased to ϕ_1 to maintain the constant power. The variations entirely satisfy the mathematical expressions in Eqs. (1.12) to (1.15). Since current and phase angle increase, the reactive power generation is also increases for the excitation rise as in (1.16).

$$VI_1 \sin\phi_1 > VI \sin\phi. \tag{1.16}$$

Similarly for constant power/torque by decreasing the excitation of the generator, reverse effects can be observed in the voltage, current, phase angle and the reactive power generation.

1.5.2 Excitation Control

The generator exciter can be a DC generator driven by the same shaft of the steam turbine that drives the main generator. There are other varieties of exciter consisting of rectifier or thyristor system receives power supply from the AC bus or alternator. Voltage regulating mechanism is used in generators to control the output of exciter to change the voltage and reactive power as per the requirement. This mechanism is known as voltage regulator. Figure 1.8 shows a block diagram of excitation system with voltage regulator.

The voltage regulator senses the generator output voltages (and sometimes the current) and then makes the move to correct excitation by changing the exciter control. Some excitation system has options for manual operation of voltage regulator. In that case, operator observes the terminal voltage and adjusts the regulator until the required output voltage is achieved. In modern technology,

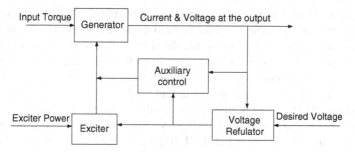

Fig. 1.8 Block diagram of generator excitation system with voltage regulator

manual system is not used unless necessary for any reason. The most important fact is that the speed of this voltage regulating system is of great interest for stability. Auxiliary control may have additional features as required for feedback of speed, frequency, acceleration, damping to prevent overshooting and many others [14]. Figure 1.9 shows two additional features: the damping transformer and the current compensator of the excitation system.

The damping transformer is an electrical dashpot device that damp out excessive action of the moving plunger. The current compensator controls the division of reactive power among generators when operate in parallel and utilize

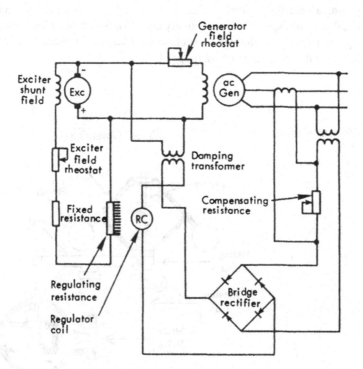

Fig. 1.9 Self-excited main exciter with Silverstat regulator (Engineers [15])

this kind of control. A voltage drop takes place in the potential circuit in relation to the line current for the current transformer and compensator resistance. The circuit connection is arranged in such a way that the voltage drop across the compensating resistance for lagging current adds to the voltage from the potential transformers. In this way, the regulator lowers the excitation voltage for the increase in lagging current (i.e. increase in reactive power output) and offers drooping characteristics so that the parallel operating generators equally share the reactive power load. There are more sophisticated excitation systems available in (Engineers [15], Chambers et al. [16], Barnes et al. [17], Report [18]) using in the growing large interconnected generators operating in parallel in the power system.

1.5.3 Governor System

Among the various prime movers, steam turbine prime mover is commonly used as maximum cases the energy is received in the form of heat that produce steam for generating massive electric power. Governor controls the amount of steam to be injected into the turbine for producing the requisite mechanical torque to the prime mover shaft. There are various types of governors. As an example, a flyball governor is shown Fig. 1.10. Two forces are acting on the flyball arms due to the shaft rotation: a centrifugal force on the masses and a downward force on the spring. The downward spring force is actually used to adjust prime mover speed by controlling the steam flow through a mechanical linkage that changes a shaft or collar position with the change of shaft speed.

Fig. 1.10 Flyball governor

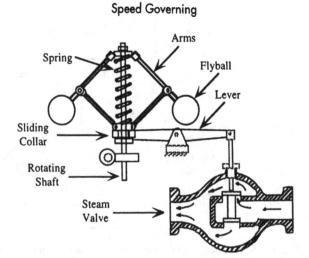

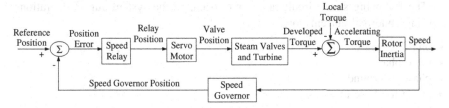

Fig. 1.11 Block diagram of steam turbine control system (Eggenberger [25])

Some steam turbine governing systems are made more efficient by using speed transducer, amplifiers comparators and force-stroke amplifiers. A block diagram of such a steam turbine control system is shown in Fig. 1.11.

The speed governor is a speed transducer which may be a mechanical, hydraulic or electrical device. It measures shaft speed and supply an output signal in any form such as position, pressure or voltage for comparison against the reference. The output of the transducer shown in the Fig. 1.11 is characteristically the position (stroke) of a rod that is proportional to the speed. The position error obtained from the comparison stroke and the preset reference position is proportional to speed. Speed relay and servomotor are used to amplify the position error and the force that controls the position error to regulate the steam and hence the torque.

By regulating the steam, the governor system on the whole changes the power at the generator output. Once the prime mover input as mechanical torque to the generator is increased, the real power (MW) at the generator output is increased. This results in the increase in the resistive component of the line current. The increased prime mover mechanical torque also increases the bus voltage and for constant power load, and this will results in decrease in the reactive power component as well as reactive component of line current and hence improve the power factor.

1.6 Protection System

Protective relays and devices are used in power system to protect electrical generators, transformers, lines and equipments against faults that may occur during normal system operation. The purpose is to quick isolation of the faulty part from the system so that remaining part can maintain the continuity of power supply. In that case, the relay operation should be selective and faster to trip the related circuit breaker. Appropriate design and operation of protection system prevent personnel injuries and damage to the equipments. It also minimizes power interruptions, effects of faults and fault-related disturbances on the system.

The following short circuit faults may occur in the system during operations causing abnormal current flow:

- phase to phase,
- three phase,
- phase to ground,
- phase to phase to ground,
- three phase to ground.

The magnitude of current depends on the impedance at the fault. If there is zero impedance at the fault, the current will be high, and for high impedance at the fault, the current will be small. The faulty section of the system is detected and isolated by the protection system. Fuses and miniature circuit breakers are used for lower voltage and small power supply in the distribution. In transmission and distribution, carrying bulk power (usually more than 1 MVA) relays are used to sense the abnormal current and initiate the circuit interrupting equipment such as circuit breaker through external control circuits or mechanically coupled mechanisms.

In distribution feeders, mostly over current and earth fault relays are used in protection system. These protections have instantaneous or time-delay character-istics. Time-delay relays can be co-ordinated with the downstream relays for correct isolation of the faulty section. For parallel feeders, directional over current and earth fault relays are used to correctly isolate the faulty feeder. Like feeder protection distribution transformers have similar over current and earth fault protection. Multiple transformers operating in parallel have reverse power relays to prevent power flow in reverse direction. Besides these over current and earth fault relays, transformers also have differential and restricted earth fault relays for selective and fast operation to isolate the faulty transformers. Differential and restricted earth fault protections are known as current balance protection which covers the zone between measuring points (HV side and LV side of transformers) of the currents that pass through the relays. These protections are sensitive for internal faults and stable for through faults (faults at outside the zone) as well as provide the high-speed operation to isolate the faulty section. Differential and time delays over current relays are also used for generator and busbar protections.

Transmission and sub-transmission lines and distribution feeders protections at voltages from 33 kV up to the highest transmission voltages are mostly made of distance relays which are operated based on the impedance of the line and the fault (if any).

The following are the main characteristics of distance relays:

- high speed (for zone 1),
- directional,
- discriminates through measuring impedance,
- time graded (zones 2 and above),
- self-contained.

The distance relay has the facility to measure the angle and the magnitude of the impedance to the fault by which it acquires the directional characteristics. The various directional impedance characteristics of the distance relays are given as follows:

- mho,
- quadrilateral,
- elliptical,
- offset mho,
- reactance.

These directional impedance characteristics are important in choosing relays for different transmission line and distribution feeder protections.

Distance relay protection scheme uses communication facility between two ends of the transmission lines for intertrip and protection signalling with various distance schemes such as trip acceleration, permissive under reach, permissive overreach, blocking scheme. This communication facility with the various distance schemes provides accurate and high-speed isolation of the faulty lines from the transmission network.

There is always some possibility of protection failure due to operation of the primary relays. Therefore, it is essential to add on backup protection besides the primary protection to ensure clearance of the fault from the system. In distribution system, the consequences of maloperation or failure to operate are less serious than in transmission systems. Hence, backup protection in distribution system can be simple and is often inherent in the main protection or watched over by the immediate upstream time graded primary relays. In transmission system (132 kV and above), where the interconnection is more complex, duplicate distance relays are used as backup protection to improve reliability. The backup and main protections should have different operating principle, so that abnormal events causing failure of the one have different influence on the other.

There are other relays which are used for special purposes such as loss of excitation for generators, under-voltage relay, over-voltage relay, power factor relay. More details about the protective relays and application are found in Blackburn [19], GeneralElectricCompany [20], Elmore [21].

1.7 SCADA System

Supervisory control and data acquisition (SCADA) system is an important infrastructure of electricity grids. Its main role in power system is to provide monitoring, control and automation functions that improve operational reliability supervisory control and data acquisition. In addition, the SCADA system is very useful for acquiring valuable knowledge and capabilities essential for the business function of utility companies in delivering power in a reliable and safe manner.

SCADA system has three essential components as given below:

- remote telemetry units (RTU),
- communications,
- human–machine interface (HMI).

The role of RTU is to collect information at a site. Communications transport that information from regional RTU sites to a central location and intermittently return commands to the RTU. The function of HMI is to display the information in graphics form, archive the receiving data, transmit alarms and allow operator control as necessary.

Communications within a plant are done by any of the following networking media:

- Ethernet cables (Cat5, Cat5e, Cat6, Cat6a),
- coaxial cable,
- token ring cables (Cat4),
- telecommunications cable (Cat2/telephone cord),
- optical fibre cable.

But radio is the most common use for communications in regional systems. The HMI is basically a PC system that runs powerful graphic and alarm software programs. Figure 1.12 shows a local area network (LAN) of substation, and Fig. 1.13 shows a SCADA system with interconnected LAN.

In this stage of technology, general-purpose computers (PC) have the capability for parallel computing and can run potential softwares that speed up problem solutions. Therefore, they can be used in computerized control system for the high-voltage and medium-voltage transformer substations. The LAN configuration shown in Fig. 1.6 can easily be housed in a substation control room.

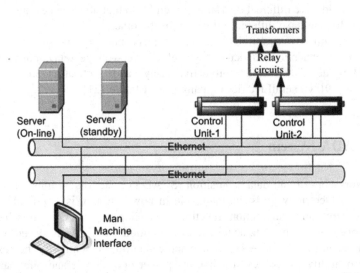

Fig. 1.12 Local area network in an electrical substation (Islam and Kamruzzaman [26])

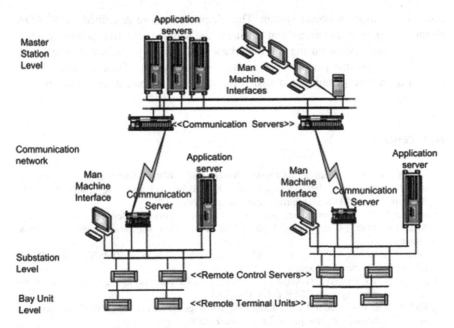

Fig. 1.13 SCADA systems with interconnected local area network (Islam and Kamruzzaman [26])

In the figures, servers (PC) that run controlling software modules are duplicated to secure the reliability of function. In the systems, one server is assigned for operating in online mode as the main control, while the other is kept in the fully operational mode for immediate take over (standby mode) as a backup control. The servers are similarly equipped with the breakdown monitoring and switchover control process. For multitask operating system, they are capable to function through Ethernet, RS485 communication interface and suitable communication protocol. SCADA system is built up with a number of servers such as application servers, communication servers and remote control servers [22–24]. The central or master station hardware architecture is built up to function at real time with the RTUs and servers distributed in different substations.

1.8 Conclusions

This chapter discussed the characteristics of the various components of a power system those are useful during normal operating conditions and during disturbances. The power system generation, transmission and distribution become very complex due to increase in the area of distribution as well as increase in power demand for rising population. Engineers have to solve enormous problem to maintain the quality and reliability of the power supply. SCADA system is the backbone of modern power system. This chapter therefore described the SCADA system along with the areas that manages the voltage, reactive power and frequency of the system during the disturbances. It will provide the engineers the ability to analyse the performance of the complex system. Their knowledge will help in operation, planning and design for future development more efficiently.

References

1. (2003) Network Planning Criteria—Power and Water Corporation, http://www.powerwater.com.au/__data/assets/pdf_file/0009/3501/network_planning_criteria_0304.pdf
2. Kimbark EW (1971) Direct current transmission. Wiley, NY
3. Uhlman E (1975) Power transmission by direct current. Springer, Berlin
4. Arrillaga J (1983) High voltage direct current transmission, IEE power engineering series, London
5. Song YH, Johns AT (1999) Flexible ac transmission systems (FACTS), IEE Power Engineering Series 30, London
6. Hingorani NG, Gyugyi L (2000) Understanding FACTS: concepts and technology of flexible AC transmission systems. IEEE Press, Piscataway
7. Rudervall R, Charpentier J et al. (2000) High voltage direct current (HVDC) transmission systems technology review paper. Energy week 2000
8. Kueck JD, Kirby BJ et al. (2004) Measurement practices for reliability and power quality. Oak Ridge National Laboratory, Oak Ridge, Tennessee 37831-6285
9. Heydt G (1991) Electric Power Quality. Stars in a Circle Publications
10. McGranaghan M, Mueller D (1885) Effects of voltage sags in process industry applications. IEEE/KTH Power Tech Conference. Stockholm, Sweden, pp 4–10
11. Bendre A, Divan D et al. (2004) Equipment failures caused by power quality disturbances. Industry applications conference, 2004. 39th IAS annual meeting. Conference record of the 2004 IEEE. vol. 1
12. Gonen T (1986) Electric power distribution system engineering. McGraw Hill, New York
13. Mithulananthan N, Salama M et al (2000) Distribution system voltage regulation and var compensation for different static load models. Int J Electr Eng Educ 37(4):384–395
14. Concordia C, Ternoshok M (1967) Generator excitation systems and power system performance. IEEE summer power meeting. Portland, Oreg: Paper 31 CP 67–536
15. Engineers CS (1950) Electrical transmission and distribution reference book. Westinghouse Electric Corporation, Pittsburgh
16. Chambers GS, Rubenstein AS et al. (1961) Recent developments in amplidyne regulator excitation systems for large generators. AIEE Trans PAS-80:1066–1072
17. Barnes HC, Oliver JA et al. (1968) Alternator-rectifier exciter for Cardinal Plant. IEEE Trans PAS-87:1189–1198
18. Report IC (1969) Proposed excitation system definitions for synchronous machines. IEEE Trans Power Apparatus Syst PAS-88(8):1248–1258
19. Blackburn JL (1987) Protective relaying—principles and practice. Marcel Dekker, Inc., New York
20. GeneralElectricCompany (1987) Protective relays application guide/ GEC measurements, Stafford, engineering: GEC Measurements, The General Electric Company (Great Britain)
21. Elmore WA (2004) Protective relaying—theory and applications. Marcel Dekker, Inc, NY
22. Gaushell DJ, Darlington HT (1987) Supervisory control and data acquisition. Proceedings of the IEEE
23. Sciacca SC, Block WR (1995) Advanced SCADA concepts. IEEE Comput Appl Power 8(1):23–28

24. Ackerman WJ (1999) Substation automation and the EMS. Proceedings IEEE transmission and distribution conference
25. Eggenberger MA (1960) A simplified analysis of the no-load stability of mechanical-hydraulic speed control systems for steam turbines, ASME Winter annual meeting New York: Paper 60-WA-34
26. Islam MF, Kamruzzaman J (2006) Implementation of ANN based Tap-changer control of transformers in transmission and distribution system. Australasian universities power engineering conference (AUPEC' 2006), Melbourne, Victoria

Chapter 2
Smart Grid

Md Rahat Hossain, Amanullah M. T. Oo
and A. B. M. Shawkat Ali

Abstract All over the world, electrical power systems are encountering radical change stimulated by the urgent need to decarbonize electricity supply, to swap aging resources and to make effective application of swiftly evolving information and communication technologies (ICTs). All of these goals converge toward one direction; 'Smart Grid.' The Smart Grid can be described as the transparent, seamless, and instantaneous two-way delivery of energy information, enabling the electricity industry to better manage energy delivery and transmission and empowering consumers to have more control over energy decisions. Basically, the vision of Smart Grid is to provide much better visibility to lower-voltage networks as well as to permit the involvement of consumers in the function of the power system, mostly through smart meters and Smart Homes. A Smart Grid incorporates the features of advanced ICTs to convey real-time information and facilitate the almost instantaneous stability of supply and demand on the electrical grid. The operational data collected by Smart Grid and its sub-systems will allow system operators to quickly recognize the best line of attack to protect against attacks, susceptibility, and so on, sourced by a variety of incidents. However, Smart Grid initially depends upon knowing and researching key performance components and developing the proper education program to equip current and future workforce with the knowledge and skills for exploitation of this greatly advanced system. The aim of this chapter is to provide a basic discussion of the Smart Grid concept, evolution and components of Smart Grid, environmental impacts of Smart Grid and then in some detail, to describe the technologies that are required for its realization. Even though the Smart Grid concept is not yet fully defined, the chapter will be helpful in describing the key enabling technologies and thus allowing the reader to play a part in the debate over the future of the Smart Grid. The chapter concludes with the experimental description and results of developing a hybrid prediction method for solar power which is applicable to successfully implement the 'Smart Grid.'

M. R. Hossain (✉) · A. M. T. Oo · A. B. M. S. Ali
Power Engineering Research Group, Central Queensland University, Rockhampton, QLD 4701, Australia
e-mail: m.hossain@cqu.edu.au

A. B. M. S. Ali (ed.), *Smart Grids*, Green Energy and Technology,
DOI: 10.1007/978-1-4471-5210-1_2, © Springer-Verlag London 2013

2.1 Introduction

The majority of the world's electricity delivery system or 'grid' was built when energy was reasonably low cost. While minor upgrading has been made to meet rising demand, the grid still operates the way it did almost 100 years ago—energy flows over the grid from central power plants to consumers, and reliability is ensured by preserving surplus capacity [1]. The result is an incompetent and environmentally extravagant system that is a foremost emitter of greenhouse gases, consumer of fossil fuels, and not well suited to distributed, renewable solar and wind energy sources. Additionally, the grid may not have ample capacity to meet future demand.

Continued economic growth and fulfillment of high standards in human life depends on reliable and affordable access to electricity. Over the past few decades, there has been a paradigm shift in the way electricity is generated, transmitted, and consumed. However, fossil fuels continue to form a dominant initial source of energy in the industrialized countries. The steady economic growth of some of those industrialized countries gradually exposed the unsustainable nature of the energy policy that is highly dependent on foreign fossil fuels. On the other hand, an aging power grid that faces new challenges posed by higher demands and increasing digital and nonlinear loads has placed new reliability concerns as observed with frequent outages in the recent years. Sensitivity of digital equipment, such as data centers, and consumer electronics, into intermittent outages has redefined the concept of reliability. As a result, power generation, transmission, and consumption has been the focus of investigations as to see what remedies will address the above challenges, thereby transforming the power grid into a more efficient, reliable, and communication-rich system [2]. Smart power grid is a host of solutions that is aimed to realize these lofty goals by empowering customers, improving the capacity of the transmission lines and distribution systems, providing information and real-time pricing between the utility and clients, and higher levels of utilization for renewable energy sources to name a few.

The present revolution in communication systems, particularly stimulated by the Internet, offers the possibility of much greater monitoring and control throughout the power system and hence more effective, flexible, and lower-cost operation. The Smart Grid is an opportunity to use new information and communication technologies (ICTs) to revolutionize the electrical power system [3]. However, due to the huge size of the power system and the scale of investment that has been made in it over the years, any significant change will be expensive and requires careful justification.

The consensus among climate scientists is clear that man-made greenhouse gases are leading to dangerous climate change. Hence, ways of using energy more effectively and generating electricity without the production of CO_2 must be found. The effective management of loads and reduction in losses and wasted energy need accurate information, while the use of large amounts of renewable generation requires the integration of the load in the operation of the power system in order to help balance supply and demand. Smart meters are an important element of the Smart Grid as they can provide information about the loads and hence the power flows throughout the network. Once all the parts of the power system are monitored, its state becomes observable and many possibilities for control emerge. In the future, the anticipated future de-carbonized electrical power system is likely to rely on generation from a combination of renewables, nuclear generators, and fossil-fuelled plants with carbon capture and storage. This combination of generation is difficult to manage as it consists of variable renewable generation and large nuclear and fossil generators with carbon capture and storage that, for technical and commercial reasons, will run mainly at constant output [3]. It is hard to see how such a power system can be operated cost-effectively without monitoring and control provided by a Smart Grid.

2.2 Smart Grid: The Definitions

The concept of Smart Grid combines a number of technologies, customer solutions and addresses several policy and regulatory drivers. Smart Grid does not have any single obvious definition. The European Technology Platform [4] defines the Smart Grid as:

> A Smart Grid is an electricity network that can intelligently integrate the actions of all users connected to it—generators, consumers and those that do both—in order to efficiently deliver sustainable, economic and secure electricity supplies.

In Smarter Grids: The Opportunity [5], the Smart Grid is defined as:

> A smart grid uses sensing, embedded processing and digital communications to enable the electricity grid to be observable (able to be measured and visualised), controllable (able to manipulated and optimised), automated (able to adapt and self-heal), fully integrated (fully interoperable with existing systems and with the capacity to incorporate a diverse set of energy sources).

According to the U.S. Department of Energy [6]:

> A smart grid uses digital technology to improve reliability, security, and efficiency (both economic and energy) of the electrical system from large generation, through the delivery systems to electricity consumers and a growing number of distributed-generation and storage resources.

From the aforementioned definitions, the Smart Grid can be described as the transparent, seamless, and instantaneous two-way delivery of energy information, enabling the electricity industry to better manage energy delivery and transmission and empowering consumers to have more control over energy decisions. A Smart Grid incorporates the benefits of advanced communications and information technologies to deliver real-time information and enable the near-instantaneous balance of supply and demand on the electrical grid. One significant difference between today's grid and the Smart Grid is two-way exchange of information between the consumer and the grid. For example, under the Smart Grid concept, a smart thermostat might receive a signal about electricity prices and respond to higher demand (and higher prices) on the grid by adjusting temperatures, saving the consumer money while maintaining comfort. Figure 2.1 shows a snapshot of the deliverance of the Smart Grid.

Introducing Smart Grid to the electrical power grid infrastructure will:

- ensure the reliability of the grid to levels never thought possible
- allow for the advancements and efficiencies yet to be envisioned
- exerting downward pressure on electricity prices
- maintain the affordability for energy consumers
- provide consumers with greater information and choice of supply
- accommodate renewable and traditional energy resources
- enable higher penetration of intermittent power generation sources
- revolutionizing not only the utility sector but the transportation sector through the integration of electrical vehicles as generation and storage devices
- finally, the Smart Grid will promote environmental quality by allowing customers to purchase cleaner, lower-carbon-emitting generation, promote a more even deployment of renewable energy sources, and allow access to more environmentally friendly central station generation. Furthermore, the Smart Grid will allow for more efficient consumer response to prices, which will reduce the need for additional fossil fuel-fired generation capacity, thereby reducing the emission of CO_2 and other pollutants.

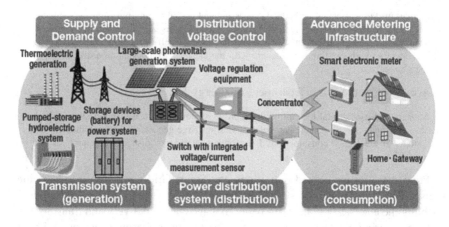

Fig. 2.1 Smart Grid [7]

2.2.1 Characteristics of Smart Grid

In short, a Smart Grid employs innovative products and services together with intelligent monitoring, control, communication, and self-healing technologies.

The literature [7–10] suggests the following attributes of the Smart Grid:

- Smart Grid allows consumers to play a part in optimizing the operation of the system and provides consumers with greater information and choice of supply. It enables demand response and demand-side management through the integration of smart meters, smart appliances and consumer loads, micro-generation, and electricity storage (electrical vehicles) and by providing customers with information related to energy use and prices. It is anticipated that customers will be provided with information and incentives to modify their consumption pattern to overcome some of the constraints in the power system.
- It better facilitates the connection and operation of generators of all sizes and technologies and accommodates intermittent generation and storage options [11]. It accommodates and facilitates all renewable energy sources, distributed generation, residential micro-generation, and storage options, thus significantly reducing the environmental impact of the whole electricity supply system. It will provide simplified interconnection similar to 'plug-and-play.'
- It optimizes and efficiently operates assets by intelligent operation of the delivery system (rerouting power, working autonomously) and pursuing efficient asset management. This includes utilizing assets depending on what is needed and when it is needed.
- It operates resiliently in disasters, physical or cyber attacks and delivers enhanced levels of reliability and security of supplying energy. It assures and improves reliability and the security of supply by anticipating and responding in a self-healing manner, and strengthening the security of supply through enhanced transfer capabilities.
- It provides power quality of the electricity supply to accommodate sensitive equipment that enhances with the digital economy.
- It opens access to the markets through increased transmission paths, aggregated supply and demand response initiatives, and ancillary service provisions.

2.2.2 Traditional Grid Versus Smart Grid

Many issues contribute to the incapability of traditional grid to competently meet the demand for consistent power supply. Table 2.1 compares the characteristics of traditional grid with the preferred Smart Grid.

Table 2.1 Comparison between the traditional and Smart Grid

Traditional grid	Smart Grid
Electromechanical, solid state	Digital/Microprocessor
One-way and local two-way communication	Global/integrated two-way communication
Centralized generation	Accommodates distributed generation
Limited protection, monitoring and control systems	WAMPAC, Adaptive protection
'Blind'	Self-monitoring
Manual restoration	Automated, 'self-healing'
Check equipment manually	Monitor equipment remotely
Limited control system contingencies	Pervasive control system
Estimated reliability	Predictive reliability

2.3 Evolution of Smart Grid

The existing electricity grid is a product of rapid urbanization and infrastructure developments in various parts of the world in the past century. Though they exist in many differing geographies, the utility companies have generally adopted similar technologies. The growth of the electrical power system, however, has been influenced by economic, political, and geographic factors that are unique to each utility company [12]. Despite such differences, the basic topology of the existing electrical power system has remained unchanged. Since its inception, the power industry has operated with clear demarcations between its generation, transmission, and distribution subsystems and thus has shaped different levels of automation, evolution, and transformation in each step.

According to Fig. 2.2, the existing electricity grid is a strictly hierarchical system in which power plants at the top of the chain ensure power delivery to customers' loads at the bottom of the chain. The system is essentially a one-way pipeline where the source has no real-time information about the service parameters of the termination points. The grid is therefore over-engineered to withstand maximum anticipated peak demand across its aggregated load. And since this peak demand is an infrequent occurrence, the system is inherently inefficient. Moreover, an unprecedented rise in demand for electrical power, coupled with lagging investments in the electrical power infrastructure, has decreased system stability [2]. With the safe margins exhausted, any unforeseen surge in demand or anomalies across the distribution network causing component failures can trigger catastrophic blackouts. To facilitate troubleshooting and upkeep of the expensive upstream assets, the utility companies have introduced various levels of command-and-control functions. A typical example is the widely deployed system known as supervisory control and data acquisition (SCADA).

Given the fact that nearly 90 % of all power outages and disturbances have their roots in the distribution network, the move toward the Smart Grid has to start at the bottom of the chain, in the distribution system. Moreover, the rapid increase in the cost of fossil fuels, coupled with the inability of utility companies to expand their generation capacity in line with the rising demand for electricity, has accelerated

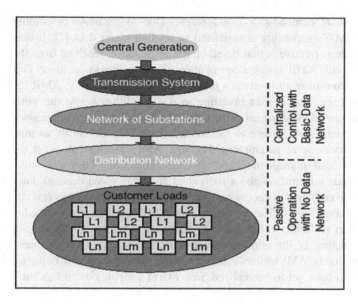

Fig. 2.2 The existing grid [12]

the need to modernize the distribution network by introducing technologies that can help with demand-side management and revenue protection.

As Fig. 2.3 shows, the metering side of the distribution system has been the focus of the most recent infrastructure investments. The earlier projects in this sector saw the introduction of automated meter reading (AMR) systems in the distribution network. AMR lets utilities read the consumption records, alarms, and status from customers' premises remotely.

Figure 2.4 suggests, although AMR technology proved to be initially attractive, utility companies have realized that AMR does not address the major issue they

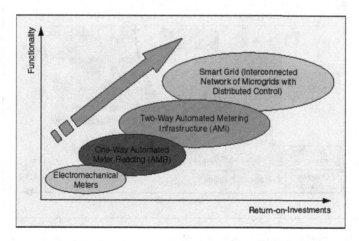

Fig. 2.3 The evolution of the Smart Grid [12]

need to solve: demand-side management. Due to its one-way communication system, AMR's capability is restricted to reading meter data [2]. It does not let utilities take corrective action based on the information received from the meters. In other words, AMR systems do not allow the transition to the Smart Grid, where pervasive control at all levels is a basic premise. Consequently, AMR technology was short-lived. Rather than investing in AMR, utilities across the world moved toward advanced metering infrastructure (AMI). AMI provides utilities with a two-way communication system to the meter, as well as the ability to modify customers' service-level parameters. Through AMI, utilities can meet their basic targets for load management and revenue protection. They not only can get instantaneous information about individual and aggregated demand, but they can also impose certain caps on consumption, as well as enact various revenue models to control their costs. The emergence of AMI heralded a concerted move by stakeholders to further refine the ever-changing concepts around the Smart Grid [2]. In fact, one of the major measurements that the utility companies apply in choosing among AMI technologies is whether or not they will be forward compatible with their yet-to-be-realized Smart Grid's topologies and technologies.

2.4 Components of Smart Grid

For the generation level of the power system, smart enhancements will extend from the technologies used to improve the stability and reliability of the generation to intelligent controls and the generation mix consisting of renewable resources.

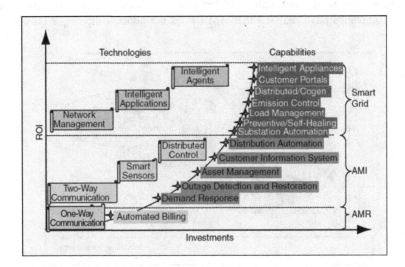

Fig. 2.4 Smart Grid returns on invesments [12]

2.4.1 Monitoring and Control Technology Component

In a conventional power system, electricity is distributed from the power plants through the transmission and distribution networks to final consumers. Transmission and distribution networks are designed to deliver the electricity at the consumer side at a predefined voltage level. Photovoltaic power generation is in general connected at the distribution level of the power system. For this reason, it is possible for the power produced by the PV to cause a 'counter' power flow from the consumer side to be delivered to other consumers through the distribution network. This phenomenon may present two challenges: an increase in the voltage in areas with high PV production and voltage fluctuation throughout the system due to the intermittency characteristics of the PV production [13]. Intelligent transmission systems include a smart intelligent network, self-monitoring and self-healing, and the adaptability and predictability of generation and demand robust enough to handle congestion, instability, and reliability issues. This new resilient grid has to resist shock (durability and reliability), and be reliable to provide real-time changes in its use. Taking these issues into consideration, voltage control systems that incorporate optimal power flow computation software are developed. These systems have been designed to rapidly analyze power flow to forecast the voltage profile on the distribution network, and, in some cases, control voltage regulation equipment to ensure the appropriate voltage. The optimal control signal is developed through optimal power flow calculation. Figure 2.5 overviews the distribution and automation system for electrical power companies.

2.4.2 Transmission Subsystem Component

The transmission system that interconnects all major substation and load centers is the backbone of an integrated power system. Transmission lines must endure dynamic changes in load and emergency without service interruptions. Strategies to achieve Smart Grid performance at the transmission level include the design of analytical tools and advanced technology with intelligence for performance analysis such as dynamic optimal power flow, robust state estimation, real-time stability assessment, and reliability and market simulation tools [13].

Real-time monitoring based on PMU, state estimators sensors, and communication technologies are the transmission subsystem's intelligent enabling tools for developing smart transmission functionality.

2.4.3 Smart Devices Interface Component

Smart devices for monitoring and control form part of the generation components' real-time information processes. These resources need to be seamlessly integrated in the operation of both centrally distributed and district energy systems. Apart

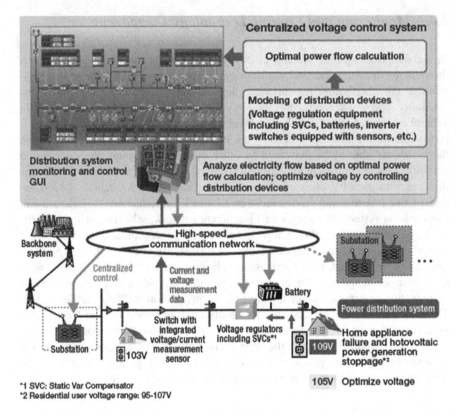

Fig. 2.5 Distribution and automation system for electrical power companies [7]

from a physical model of a smart device, there is also a need for a logical model for a smart device. Such a model must outline what a smart device offers to a smart space with regard to the services it can provide to the environment. The model must also outline interactions between smart devices, changes in the state of smart device operation, and smart services within a smart space. There are various models present today that have similar approaches to modeling device [13]. Two such models would include Home Plug and Play (HPnP), which is slightly out-of-date but still applicable, and the newer Universal Plug and Play (UPnP), which is an open standards body. These two standards bodies have modeled devices by the services that they offer and have also developed interaction models for device communication. Another emerging standard for defining services in an abstract way is with the use of the Web Services Definition Language (WSDL). Along with describing services a smart device can provide, there must also be a way of representing changes in states of smart devices and how devices react to these changes within a smart environment.

2.4.4 Intelligent Grid Distribution Subsystem Component

The distribution system is the final stage in the transmission of power to end users. At the distribution level, intelligent support schemes will have monitoring capabilities for automation using smart meters, communication links between consumers and utility control, energy management components, and AMI [13]. The automation function will be equipped with self-learning capability, including modules for fault detection, voltage optimization and load transfer, automatic billing, restoration and feeder reconfiguration, and real-time pricing. Electrical companies are accelerating efforts to develop an advanced meter infrastructure (AMI) to improve customer services and reduce meter reading costs. An essential element in this AMI is the smart meter. A smart meter is a device that not only measures the electricity consumption but also able to communicate with a center. Developing the communication network between the meter and the center presents several challenges, including costs and reliability. AMI technologies and systems need to be developed to ensure reliability and flexibility in measuring and controlling electricity meters through next-generation wireless mesh networks [13]. Wireless mesh networks provide a transmission method that links electrical meters to relay data by each meter through other meters, using a multi-hop network scheme. This network is helpful reduce the time required to acquire data while at the same time curtailing costs. While wireless mesh networks present cost benefits, some challenges have to be overcome to ensure practical application. Simultaneous transfer of data between meters at the same frequency can cause signal collision, preventing reliable data collection. Figure 2.6 represents the advanced metering infrastructure those are being used in present days for electrical power companies.

2.4.5 Storage Component

Due to the unpredictability of renewable energy and the disjoint between peak availability and peak consumption, it is important to find ways to store the generated energy for later on use. Options for energy storage technologies include pumped hydro, advance batteries, flow batteries, compressed air, super-conducting magnetic energy storage, super-capacitors, and flywheels [13]. Associated market mechanism for handling renewable energy resources, distributed generation, environmental impact, and pollution has to be introduced in the design of Smart Grid component at the generation level.

2.4.6 Demand-side Management Component

Demand-side management (DSM) and energy efficiency options were developed for effective means of modifying the customer demand to cut operating expenses

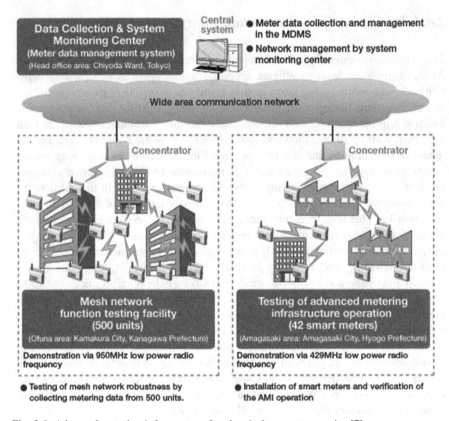

Fig. 2.6 Advanced metering infrastructure for electrical power companies [7]

from expensive generators and suspend capacity addition [13]. DSM options provide reduced emissions in fuel production, lower costs, and contribute to reliability of generation. These options have an overall impact on the utility load curve. Electrical power companies are obligated to maintain constant frequency levels and the instantaneous balance between demand and supply by adjusting output through the use of thermoelectric and pumped storage generation. With the expected increase in photovoltaic power generation, the supply power may fluctuate considerably due to changes in the weather. Imbalance between demand and supply causes fluctuation in the system frequency that may, in turn, affect negatively user appliances and, in a worst case, lead to a power outage [13]. In order to resolve this issue, optimal demand–supply control technologies are required to develop to control not only conventional generators but also batteries and other storage devices. Figure 2.7 demonstrates the advanced demand and supply planning and control system for electrical power companies and transmission system operators.

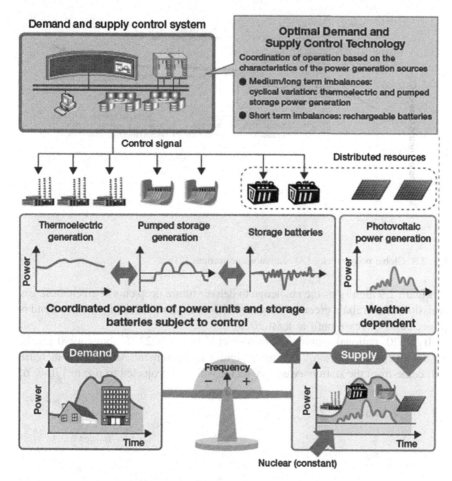

Fig. 2.7 Demand and supply planning and control system for electrical power companies and transmission system operators [7]

2.5 The Environmental Impacts of Smart Grid

The resulting forecasts of global power sector CO_2 emissions are illustrated in Fig. 2.8. The conservative scenario leads to 5 % reduction in annual power sector CO_2 emissions by 2030, with the average annual growth rate in CO_2 emissions dropping from 0.7 to 0.5 % [2]. The expanded scenario produces even further reductions. Power sector CO_2 emissions in 2030 drop by 16 % relative to the business-as-usual (BAU) case. CO_2 emissions are essentially flattened under this scenario, with the annual change in CO_2 emissions becoming an average decrease of 0.1 % per year.

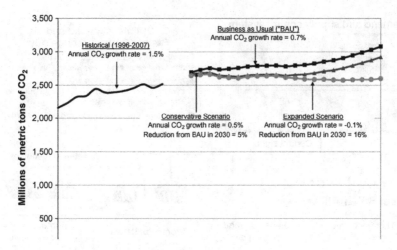

Fig. 2.8 Global power sector CO_2 emissions projections [2]

Figure 2.9 highlights the challenge to deliver future deep cuts in greenhouse gas emissions. Australia's green house gas (GHG) emissions are rising and this trend is projected to increase until at least 2020.

By 2020, national emissions are projected to reach 22 % above 1990 levels, even with current measures delivering significant abatement. Most of this increase will come from the stationary energy sector which is projected to rise to 170 % of

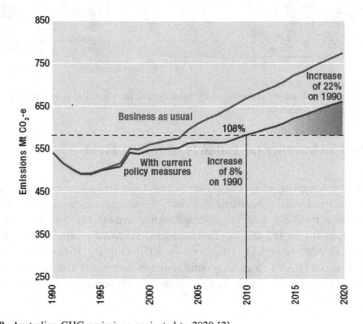

Fig. 2.9 Australian GHG emissions projected to 2020 [2]

1990 levels by 2020 [2]. These reductions are the product of several changes to the power system. Fewer coal and natural gas plants are built, because there is generally a lower need for new capacity due to the decreased demand for electricity. In the expanded scenario, much of this capacity is displaced with cleaner renewable resources. The reduction in line losses also reduces the amount of electricity that must be produced by power plants in order to meet demand.

2.5.1 Reduce Greenhouse Gas Emissions

Worldwide demand for electrical energy is expected to rise 82 % by 2030 (Energy Information Administration, US [14]). Unless revolutionary new fuels are developed, this demand will be met primarily by building new coal, nuclear, and natural gas electricity generation plants. Not surprisingly, world CO_2 emissions are estimated to rise by 59 % by 2030 as a result. The Smart Grid can help offset the increase in CO_2 emissions by slowing the growth in demand for electricity.

- Enable consumers to manage their own energy consumption through dashboards and electronic energy advisories. More accurate and timely information on electricity pricing will encourage consumers to adopt load-shedding and load-shifting solutions that actively monitor and control energy consumed by appliances.
- In deregulated markets, allow consumers to use information to shift dynamically between competing energy providers based on desired variables including energy cost, greenhouse gas emissions, and social goals. Users could include utility companies, homeowners with rooftop solar panels, and governments with landfills that reclaim methane gas. This open market approach could accelerate profitability and speed further investments in renewable energy generation [1].
- Broadcast demand–response alerts to lower peak energy demand and reduce the need for utility companies to start reserve generators. Remote energy management services and energy control operations will also advise consumers, giving them the choice to control their homes remotely to reduce energy use.
- Allow utility companies to increase their focus on 'Save-a-Watt' or 'Mega-Watt' programs instead of producing only power. These programs are effective because offsetting a watt of demand through energy efficiency can be more cost-effective and CO_2-efficient than generating an extra watt of electricity [15].

2.6 Overview of the Technologies Required for Smart Grid

To accomplish the diverse necessities of the Smart Grid, the following enabling technologies must be developed and implemented:

Sensing, measurement, control, and automation technologies: These include

- Phasor measurement units (PMU) and wide area monitoring, protection and control (WAMPAC) to ensure the security of the power system [3].
- Intelligent electronic devices (IED) to provide advanced protective relaying, measurements, fault records, and event records for the power system, integrated sensors, measurements, control and automation systems, and information and communication technologies to provide rapid diagnosis and timely response to any event in different parts of the power system [3]. These will support enhanced asset management and efficient operation of power system components, to help relieve congestion in transmission and distribution circuits and to prevent or minimize potential outages and enable working autonomously when conditions require quick resolution.
- smart appliances, communication, controls, and monitors to maximize safety, comfort, convenience, and energy savings of homes.
- smart meters, communication, displays, and associated software to allow consumers to have better choice and control over electricity use. Those will provide consumers with accurate bills, accurate real-time information on their electricity use, and enable demand management and demand-side participation.

Information and communications technologies: These include

- two-way communication technologies to provide connectivity between different components in the power system and loads [3].
- open architectures for plug-and-play of home appliances; electrical vehicles; and micro-generation.
- communications and the necessary software and hardware to provide customers with greater information enable customers to trade in energy markets.
- software to ensure and maintain the security of information and standards to provide scalability and interoperability of information and communication systems [3].

Power electronics and energy storage: These include

- High-voltage DC (HVDC) transmission and back-to-back schemes and flexible AC transmission systems (FACTS) to enable long-distance transport and integration of renewable energy sources [3].
- different power electronic interfaces and power electronic supporting devices to provide efficient connection of renewable energy sources and energy storage devices.
- series capacitors, unified power flow controllers (UPFC) and other FACTS devices to provide greater control over power flows in the AC grid [3].
- HVDC, FACTS, and active filters together with integrated communication and control to ensure greater system flexibility, supply reliability, and power quality [3].
- power electronic interfaces and integrated communication and control to support system operations by controlling renewable energy sources, energy storage, and consumer loads.
- energy storage to facilitate greater flexibility and reliability of the power system.

2.7 The Future: The Key Challenges

- Strengthening the grid—ensuring that there is sufficient transmission capacity to interconnect energy resources, specially renewable resources, across Europe [2]
- Moving offshore—developing the most efficient connections for offshore wind farms and for other marine technologies
- Developing decentralized architectures—enabling smaller-scale electricity supply systems to operate harmoniously with the total system
- Communications—delivering the communications infrastructure to allow potentially millions of parties to operate and trade in the single market [2]
- Active demand side—enabling all consumers, with or without their own generation, to play an active role in the operation of the system [2]
- Integrating intermittent generation—finding the best ways of integrating intermittent generation including residential micro-generation [2]
- Enhanced intelligence of generations
- Capturing the benefits of distributed generation and storage [2]
- Preparing for electrical vehicles—whereas Smart Grids must accommodate the needs of all consumers, electrical vehicles are particularly emphasized due to their mobile and highly dispersed character and possible massive deployment in the next years, which would yield a major challenge [2].

2.8 Experiments to Select the Base Regression Algorithms of the Hybrid Prediction Method for Smart Grid

Computational intelligence (CI) holds the key to the development of Smart Grid to overcome the challenges of planning and optimization through accurate prediction of renewable energy sources (RES), managing data and communications, control and protection of power plants. It was observed that the hybrid prediction method or ensemble model (i.e. a set of different regression algorithms or machine learning techniques whose discrete predictions are united to generate an ultimate aggregated prediction [16]) is suitable for a reliable Smart Grid energy management. The following experiments investigate the applicability of heterogeneous regression algorithms for 6-h ahead solar power hybrid prediction using historical data of Rockhampton, Australia. Prediction reliability of the proposed hybrid prediction method is carried out in terms of error validation metrics such as *Correlation Coefficient (CC)*, *Mean Absolute Error (MAE)*, *Mean Absolute Percentage Error (MAPE)*, *Root Mean Squared Error (RMSE)*, *Root Mean Squared Percentage Error (RMSP)*, *Mean Absolute Scaled Error (MASE)*, *Root Relative Squared Error (RRSE)* and *Relative Absolute Error (RAE)*. The experimental results show that the proposed hybrid method achieved acceptable prediction accuracy. This potential

Table 2.2 Results of applying 10-fold cross-validation method on the data set

		CC	RMSE	MAE	RRSE	RAE
10-Fold cross-validation	LR	0.89	150.14	66.35	46.30	27.44
	RBF	0.13	321.58	240.09	99.16	99.30
	SVM	0.88	164.18	46.58	50.63	19.26
	MLP	0.99	14.81	9.74	4.57	4.03
	PR	0.89	150.14	66.31	46.30	27.42
	SLR	0.87	158.56	83.15	48.89	34.39
	LMS	0.88	165.15	47.94	50.92	19.83
	AR	0.94	108.94	80.99	33.59	33.50
	LWL	0.81	190.26	146.09	58.67	60.42
	IbK	0.93	124.41	90.86	38.36	37.58

hybrid model is applicable as a local predictor for any proposed hybrid method in real-life application for 6 h in advance prediction to ensure constant solar power supply in the Smart Grid operation.

2.8.1 Experiment Design

Ten popular regression algorithms namely Linear Regression (LR), Radial Basis Function (RBF), Support Vector Machine (SVM), Multilayer Perceptron (MLP), Pace Regression (PR), Simple Linear Regression (SLR), Least Median Square (LMS), Additive Regression (AR), Locally Weighted Learning (LWL), and Instance Based K-nearest neighborhood (IbK) Regression have been used to find out ensemble generation. A unified platform is used with WEKA release 3.7.3 for all of the experiments. The WEKA 3.7.3 Developer Version is a Java-based learning tool and data mining software which is issued under the GNU General Public License [17]. WEKA is an efficient data pre-processing tool which encompasses a comprehensive set of learning algorithms with graphical user interface as well as command prompt. Regression, classification, and association rule mining, clustering, and attribute selection all are integrated in WEKA.

To estimate model accuracy precisely, the wide-ranging practice is to perform some sort of cross-validation method as well as training and testing method for error estimation. For this chapter, both the 10-fold cross-validation method and training (70 %) and testing (30 %) method are exercised. In Table 2.2, the results of applying 10-fold cross-validation method on initially selected regression algorithms are demonstrated.

The above results obtained from applying 10-fold cross-validation method on the data set clearly show that in terms of the *MAE*, the most accurate one is the MLP regression algorithm. Next to the MLP, SVM is in the second best position and LMS regression algorithm is in the third best position. In Table 2.3, the results of applying training and testing error estimator method on initially selected regression algorithms are illustrated.

Table 2.3 Results of applying training and testing method on the data set

		CC	RMSE	MAE	RRSE	RAE
Training (70 %) and testing (30 %)	LR	0.88	150.84	66.91	47.21	27.75
	RBF	0.12	317.38	239.64	99.34	99.40
	SVM	0.87	165.56	47.19	51.82	19.58
	MLP	0.99	16.90	11.73	5.29	4.87
	PR	0.88	150.88	66.65	47.22	27.65
	SLR	0.87	159.22	82.74	49.83	34.32
	LMS	0.87	164.69	48.69	51.55	20.19
	AR	0.93	114.90	83.97	35.96	34.83
	LWL	0.80	190.39	146.48	59.59	60.76
	IbK	0.92	128.82	94.23	40.32	39.09

Table 2.4 Six-hour ahead prediction errors of different regression algorithms

Six hours in advance prediction error summary

	CC	MAE	MAPE	RMSE	RMSPE	MASE	RANK
LMS	0.96	77.19	17.65	107.94	29.19	0.63	1
MLP	0.99	91.02	20.17	119.73	31.62	0.74	2
SVM	0.96	126.88	21.72	135.15	24.01	1.03	3
LR	0.96	148.41	24.07	155.82	25.12	1.21	4
LWL	-0.15	213.33	46.86	271.22	72.39	1.73	5
IbK	0.88	275.00	47.90	290.18	53.66	2.24	6
AR	0.93	298.45	48.21	306.03	49.05	2.43	7

The above results obtained from applying training (70 %) and testing (30 %) method on the data set clearly show that in terms of the *MAE,* the most accurate one is the MLP regression algorithm. Next to the MLP, SVM is in the second best position and LMS regression algorithm is in the third best position as well. Both the 10-fold cross-validation and training and testing methods suggest the top-most three accurate and potential regression algorithms for this research are MLP, SVM, and LMS in descending order. Afterward, six-hour ahead solar radiation prediction with the potential regression algorithms were performed to compare the errors of the individual prediction to select three decisive regression algorithms for the ensemble generation. In Table 2.4, the summary of six hours in advance prediction error for different regression algorithms with various prediction accuracy validation metrics is presented.

From the individual prediction results, the regression algorithms are ranked. According to [46], MAE is strongly suggested for error measurement. Hence, the ranking is done based on the *Mean Absolute Error (MAE)* of those regression algorithms' predictions. Based on the *MAE* and the *MASE* the top-ranked three regression algorithms for ensemble generation are LMS, MLP, and SVM. In Figs. 2.10, 2.11, 2.12, the comparison between the actual and predicted values of the LMS, MLP, and SVM regression algorithms is graphically presented. In all

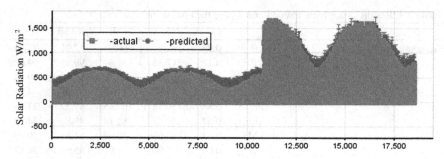

Fig. 2.10 Prediction performance of LMS

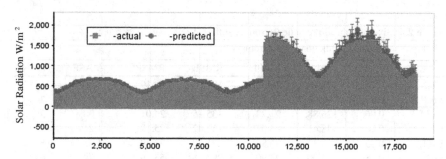

Fig. 2.11 Prediction performance of MLP

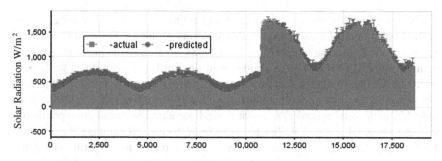

Fig. 2.12 Prediction performance of SVM

those figures, x axis represents the number of instances and y axis represents the solar radiation measured in W/m^2.

Sophisticated intelligent techniques are mandatory to handle the Smart Grid operation in an efficient and economical way. The strengths of CI paradigms have been demonstrated to resolve the ensemble generation confront for the proposed hybrid method for solar power prediction. Such hybrid solar power prediction methods are promising solutions to convey the expectations of a Smart Grid.

2.9 Summary

The aim of this chapter is to provide a basic discussion on the background of the Smart Grid; its concept and definition. Even though the Smart Grid concept is not yet fully defined, a working definition of the Smart Grid was given. Characteristics of Smart Grid and comparison between traditional or existing grid and Smart Grid are also included in this chapter. Afterward the history of the evolution of Smart Grid and the specific components of prospective Smart Grid function were provided. Environmental impact of implementing Smart Grid, particularly the way it can be used to reduce greenhouse gas emissions, and the overview of the technologies required for Smart Grid are discussed in this chapter. Directions of some future key challenges those need to be resolved for the successful implementation of Smart Grid are also depicted in this chapter. The chapter concludes with the experimental description and results of developing a hybrid prediction method for solar power which is applicable to successfully implement the 'Smart Grid.'

References

1. Frye W (2008) Transforming the electricity system to meet future demand and reduce greenhouse gas emissions. Cisco Internet Business Solutions Group
2. Hossain MR, Amanullah MT, Ali S (2010) Evolution of smart grid and some pertinent issues. In: 20th Australasian Universities power engineering conference (AUPEC) available via Online. http://www.ieeeexplore.com/stamp/stamp.jsp?tp=&arnumber=5710797&isnumber=5710678. Cited 09 Feb 2013
3. Ekanayake J, Liyanage K, Wu J et al (2012) Smart grid, technology and applications. Wiley, NY
4. European Commission (2006) European smart grids technology platform: vision and strategy for Europe's electricity. Available via Online. http://www.ec.europa.eu/. Cited 02 Feb 2013
5. Department of Energy and Climate Change, UK (2009) Smarter grids: the opportunity. Available via Online. http://www.decc.gov.uk/. Cited 01 Feb 2013
6. Department of Energy, U.S. (2009) Smart grid system report. Available via Online. http://www.doe.energy.gov/. Cited 30 Jan 2013
7. Mitsubishi Electric (2013) Environmental technology R&D achievements. Available via Online. http://www.mitsubishielectric.com/company/environment/report/products/randd/smartgrid/. Cited 06 Feb 2013
8. National Energy Technology Laboratory US (2009) A compendium of modern grid technologies. Available via Online. http://www.netl.doe.gov/. Cited 06 Feb 2013
9. European Commission (2009) ICT for a low carbon economy. Available via Online. http://www.ec.europa.eu/. Cited 17 Jan 2013
10. World Economic Forum (2009) Accelerating smart grid investments. Available via Online. http://www.weforum.org/pdf/SlimCity/SmartGrid2009.pdf. Cited 07 Feb 2013
11. Harris A (2009) Smart grid thinking. Eng Technol 4 (9):46-49. Available via Online. http://www.ieeexplore.ieee.org/stamp/stamp.jsp?tp=&arnumber=5160830&isnumber=5153158. Cited 09 Jan 2013
12. IEEE Power and Energy Magazine (2010) The path of the smart grid. Available via Online. http://www.ieeexplore.ieee.org/stamp/stamp.jsp?arnumber=05357331. Cited 05 Jan 2013
13. Momoh J (2012) Smart grid: fundamentals of design and analysis. Wiley, NY

14. U. S. Energy Information Administration (2011) International energy outlook 2011. Available via Online. http://www.eia.doe.gov/oiaf/ieo/highlights.html. Cited 01 Feb 2013
15. Rogers J (2007) Utility's idea: higher bills for less electricity. Available via Online. http:// www.usnews.com/usnews/biztech/articles/070508/8energy.htm. Cited 05 Feb 2013
16. Dietterich TG (2001) Ensemble methods in machine learning multiple classifier systems. Lect Notes Comput Sci 1857:1–15
17. Bouckaert Remco R, Frank E et al (2010) WEKA manual for version 3–7-3. The University of Waikato, New Zealand

Chapter 3
Renewable Energy Integration: Opportunities and Challenges

G. M. Shafiullah, Amanullah M. T. Oo, A. B. M. Shawkat Ali, Peter Wolfs and Mohammad T. Arif

Abstract Renewable energy (RE) is staring to be used as the panacea for solving current climate change or global warming threats. Therefore, government, utilities and research communities are working together to integrate large-scale RE into the power grid. However, there are a number of potential challenges in integrating RE with the existing grid. The major potential challenges are as follows: unpredictable power generation, week grid system and impacts on power quality (PQ) and reliability. This chapter investigates the potential challenges in integrating RE as well as distributed energy resources (DERs) with the smart power grid including the possible deployment issues for a sustainable future both nationally and internationally. Initially, the prospects of RE with their possible deployment issues were investigated. Later, a prediction model was proposed that informs the typical variation in energy generation as well as effect on grid integration using regression algorithms. This chapter also investigates the potential challenges in integrating RE into the grid through experimental and simulation analyses.

3.1 Introduction

Growing concerns about energy security, energy cost and climate change have intensified the interest in harnessing energy from renewable sources. Conventional stationary energy sector consumes mostly coal, petroleum oil and natural gas (methane) as fuel to generate required electricity demand. However, a significant amount of electricity is generated from nuclear energy in few developed countries. At present, stationary energy sector is the major contributor of greenhouse gas (GHG) emission and coal-fired power stations made Australia one of the highest

G. M. Shafiullah (✉) · A. M. T. Oo · A. B. M. S. Ali · P. Wolfs · M. T. Arif
Central Queensland University, Bruce Highway, Rockhampton QLD-4702, Australia
e-mail: g.shafiullah@cqu.edu.au

A. B. M. S. Ali (ed.), *Smart Grids*, Green Energy and Technology,
DOI: 10.1007/978-1-4471-5210-1_3, © Springer-Verlag London 2013

per capita GHG emitting countries. Currently, renewable energy (RE) fulfils 15–20 % of world's total energy demand. Therefore, to reduce the GHG emission and to get energy from naturally free sources, it is urgent to maximise the RE utilisation by bringing higher percentage of RE into the national energy mix. Among different RE applications, hydroelectricity is matured and most exploited although it is strictly location-specific application. Other most promising sources of RE are wind and solar, and these two are the largely installed distributed energy generators in the world.

To integrate large amount of electricity from these sources into the grid, main concerns are the variability in energy from these sources. Traditionally, electricity generates from large generating stations which then transmitted to the distribution network and finally to the end load. Introduction of distributed generation (DG) changed this scenario as many distributed energy resources (DERs) are to be installed at the low-voltage (LV) distribution network (DN). The nature of DERs connected near the load introduces both-way power flows at the point of common coupling (PCC) in the network. Large-scale integration of such DERs can influence system reliability and power quality (PQ).

Efficient, time forward prediction can help in effective integration into the grid for better load management as RE sources are intermittent in nature. Moreover, RE particularly solar and wind potential varies with space and time; therefore, it is vital to identify the suitable location for such installation. However, the characteristics and frequent variability in RE can have serious impacts on load where energy storage can help in load management and fluctuation reduction. Therefore, this chapter describes the potential of RE, integration into the grid and impacts on the grid due to integration of these RE sources.

3.2 Current Power System

Currently, more than 80 % of the world's energy is produced from fossil fuel that pollutes surrounding environments each and every day, which causes global warming. According to the report of the international energy agency (IEA), the world electricity generation was 14,781 billion kWh in 2003 and is projected to be 21,699 and 30,116 billion kWh in 2015 and 2030, respectively, an average increase rate of 2.7 % annually [1]. GHG emissions from electricity generation are approximately 40 % of total emissions as most of that industry uses fossil fuels, particularly coal and oil, and hence are a leading contributor to global energy-related CO_2 emissions [1, 2].

The major energy resource productions in Australia are coal, uranium and natural gas. In 2007–2008, Australia's energy production (including exports) was dominated by coal, which accounted for 54 % of total Australian energy production. Over the past 10 years, from 1997–1998 to 2007–2008, energy production in Australia increased at an average rate of 3.5 % a year, compared with 3.2 % over the past 10 years [3]. Australian primary energy consumption consists predominantly of

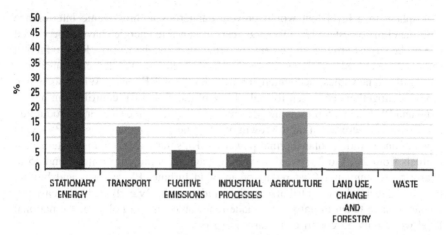

Fig. 3.1 Total Australian emissions by sector [4]

petroleum and coal. There are numerous environmental issues associated with fossil fuel extraction and usage—resource depletion, land damage and sterilisation at the extraction site, pollution during transport, smoke and smog production and acid rain [4]. It is seen that the major source of emissions is stationary energy production and the bulk of these are due to the burning of fossil fuels, in particular coal for electricity production as shown in Fig. 3.1 [4]. Australia's abundance of coal imposes environmental costs in the form of GHG, including 200 million tons of carbon dioxide equivalents (CO_2-e) released from the energy sector in 2008, more than a third of Australia's total CO_2-e emissions [3].

There is an urgent need worldwide as well as Australia to search for alternative energy sources which are free from GHG emissions. RE offers alternative sources of energy which are in general pollution free, climate friendly, unlimited, technologically effective and environmentally sustainable and started to be used as remedy for solving global warming worldwide. Therefore, an essential new research direction is to deploy large-scale RE into the energy mix for a sustainable climate-friendly environment.

3.3 Renewable Energy

Advantages of RE sources are enormous as they are free from GHGs and related global warming effects. RE is defined as an inexhaustible and sustainable energy source, and particularly in this modern environment, it is associated with climate change initiatives [5, 6]. Therefore, policy makers, power system planners, researchers and power utilities are working together worldwide to reduce GHG emissions, and hence, in 1997, a treaty was formulated called the Kyoto Protocol [7]. The objective of the Kyoto Protocol is to reduce GHG emissions into the atmosphere to a level that would prevent dangerous anthropogenic interference with the climate system [7]. Over the years, renewable energies have experienced one of the largest growths in percentage terms. In 2009, the world's RE production share has been calculated as 19.46 %. Hydroelectric is the largest contributor among the renewables, accounted for over 83 % of RE share. Wind, solar and biomass altogether accounted for only 15 % of the global RE contribution; hence, wind and PV (the most promising RE sources) have still modest energy production [8]. However, there was a rapid growth of RE generation including solar PV, wind power, concentrating solar thermal power (CSP), solar water heating systems and biofuels from 2005 to 2010 and grew at average rates ranging from around 15 to nearly 50 % annually. Cost reduction in wind turbine, PV systems and biofuel processing technology contributed to the rapid growth. By early 2011, at least 119 countries have taken initiative and made renewable support policy at the national level while only 55 countries in early 2005 [9].

Australian production of RE is dominated by bagasse, wood and wood waste and hydroelectricity, which altogether accounted for 87 % of RE production in 2007–2008. Wind, solar, and biofuels accounted for the remainder of Australia's RE production. Solar energy is mostly used for residential water heating, and this accounts for 1.5 % of final energy consumption in the residential sector. RE production increased by 6 % in the 5 years from 2002–2003 to 2007–2008 and increased by 3 % from 2006–2007 to 2007–2008 [3].

A range of policy measures have been introduced in Australia to increase electricity generation from RE sources to achieve the national goal of introducing 20–25 % RE into the energy mix. The RE sources that have experienced the greatest growth under the Australian governments' mandatory renewable energy target (MRET) are wind energy and solar energy. At the end of October 2009, there were 9 renewable electricity projects at an advanced planning stage and a further 80 projects at a less advanced stage; of these, 8 are advanced wind energy projects and 71 are wind energy projects at a less advanced stage [3]. There are 5 proposed solar energy projects in Australia, the largest of which is an 80 megawatt solar plant at Whyalla, South Australia. The largest hydroelectric power scheme in Australia is the Snowy Mountains Scheme that generates about 50 % of Australia's hydroelectric power. There is one geothermal project in operation in Australia at Birdsville, Queensland [3].

Recently, the Australian government has taken clean energy initiative (CEI) for the deployment of a range of renewable and clean energy technologies that includes [10]

- Carbon capture and storage (CCS) initiatives: CCS flagships programme accelerates the deployment of large-scale integrated CCS projects in Australia that will reduce emissions from coal use.
- Solar flagships: Solar flagships programme supports the construction and demonstration of large-scale, grid-connected solar systems in Australia that play a key role in electricity generation.

- Australian Solar Institute (ASI): ASI provides solar research and development facility and ensures collaboration with researchers in universities, institutions and industry.
- Australian Centre for Renewable Energy (ACRE): ACRE is promoting the development, commercialisation and deployment of RE technologies.
- Renewable Energy Future Fund (REFF): REFF supports the development and deployment of large- and small-scale RE projects.

In Australia, the Intelligent Grid Program [11, 12] was launched on 19 August 2008, being established under the CSIRO's Energy Transformed Flagship, and it focuses on the national need to reduce GHG emissions. Based on the focus of Australian Government Energy Policy, this study has concentrated mostly on large-scale integration of wind and solar energy into the grid.

3.3.1 Solar Energy

Solar energy is the most readily available and free source of energy since prehistoric times. Grid-connected Photovoltaic (PV) systems and DG offer various advantages over conventional generation by providing more effective utilisation of generated power. Increased penetration of PV– DG must also maintain utility grid reliability. Major components of a grid-connected PV system are shown in Fig. 3.2 [13].

An inverter system is required to transform the DC voltage produced from PV arrays. The fundamental requirements of an inverter are to limit the harmonic distortion and to ensure a constant output voltage. Sine wave inverters are mostly used for grid-connected PV systems due to the power output, system efficiencies and harmonic distortion limit [13, 14].

Currently, solar photovoltaic is the fastest growing power generation technology and an estimated 17 GW of PV was added worldwide in 2010, which was less than 7.3 GW in 2009. The annual average growth rate of solar PV exceeded 49 %, over 2005–2010 periods, which is shown in Fig. 3.3 [9].

Compared with other parts of the world, Australia is one of the best locations for solar energy as it has huge open lands and longer periods of sunshine. A total of 837 MW of PV was installed in Australia in 2011, more than twice the installed capacity added in 2010. Australia's total PV capacity has increased significantly

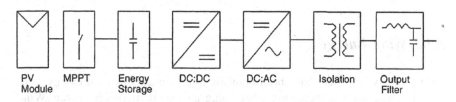

Fig. 3.2 Connection diagram for a grid-connected PV system [13]

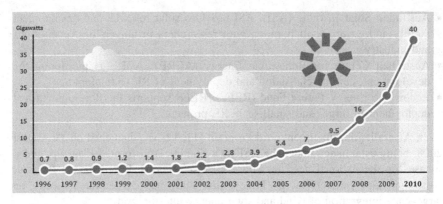

Fig. 3.3 Existing world capacity of solar PV [9]

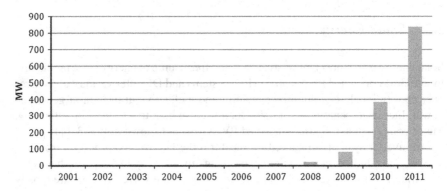

Fig. 3.4 Annual Australian PV installations 2001–2011 [15]

over the last decade and in particular over the last 2 years as shown in Fig. 3.4 [15]. This is due to a combination of factors: government renewable energy target (RET) to get 20 % electricity or 45,000 GWh from RE, solar homes and communities plan (SHCP), solar flagships, greater public awareness, feed-in-tariff, a drop in the price of PV systems, a strong Australian dollar and highly effective marketing by PV retailers [9, 15].

Solar PV systems are expected to play a promising role as a green energy source in meeting future electricity demands to build an environment-friendly sustainable power system.

3.3.2 Wind Energy

Over recent years, there have been dramatic improvements in wind energy technologies, and wind is increasingly becoming an important energy source. Wind energy can be exploited in many parts of the world, but is the most cost-effective in

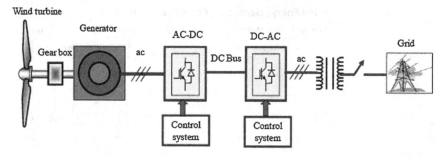

Fig. 3.5 Grid-connected wind energy system [17]

windy climates, where average wind speeds exceed 6.5 m/s. Winds are caused due to the temperature variation in the earth's surface and in the atmosphere, and in the rotation of the earth about its axis and its motion around the Sun [16].

The wind farm is composed of several wind turbines which have basic electrical components: an aerodynamic rotor, a mechanical transmission system, an electric generator, a control system, limited reactive power compensation and a step-up transformer as shown in Fig. 3.5. The generator is used for converting the mechanical power obtained from the wind turbine to electrical power. A wind turbine comprises rotor/blades for conversion of wind energy into rotational shaft energy, a nacelle with drive train that contains the generator and gear box, a tower that supports the rotor and drive train and the necessary electric equipment for connection to the grid. The majority of wind turbines offered today is of the three-bladed upwind horizontal axis type and installations intended to connect at the PCC at medium or high voltage [16, 17] of the network.

Power production from wind turbines is dependent on wind speed (v), air density (ρ) and the rotor swept area (A). Therefore, the maximum power P available from the wind can be represented as Eq. (3.1) [18].

$$P = \frac{1}{2}\rho A v^3 \tag{3.1}$$

However, the actual amount of energy production will be less as it is not possible to extract all available energy by the turbine, and therefore, a power coefficient (Cp) is defined. The ideal or maximum theoretical efficiency Cp of a turbine is the ratio of maximum power obtained from the wind to the total available power in the wind. The factor Cp = 0.593 is known as Betz coefficient or limit [18, 19]. From Eq. (3.1), it is seen that wind speed has a significant role in the amount of energy that can be produced from a wind source.

Wind energy is the fastest emerging energy technology, and total cumulative installed capacity of wind energy in the world by 2000 was 17,400 MW, while in 2011, the cumulative installed capacity is 237,669 MW as shown in Fig. 3.6. Annual installed wind capacity in 2000 was only 3,760 MW; with rapid growth, annual installed capacity in 2011 was 40,564 MW. In 2010, the rate of increase in wind energy generation globally was 24.1 %, though there had been a slight

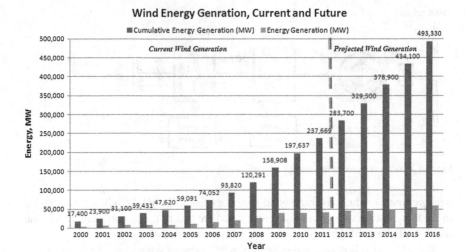

Fig. 3.6 Global wind energy installed capacity, current and projected [21]

decrease in the growth rate than earlier due to the worldwide financial crisis [20]. According to the Global Wind Energy Council, the growth rate of wind energy will increase rapidly, and over the 5 years to 2016, global wind capacity will rise to 493 GW from the 237 GW available at the end of 2011 as shown in Fig. 3.6 [21]. While the capacity installed in 2011 was 40.6 GW, the capacity predicted to be installed in 2016 is 59.24 GW; hence, the average projected annual growth rates during this period will be 13.65 % [20].

Australia has been slow to adopt wind energy to the extent that Europe has; however, in Australia, there are several large wind farms that have been commissioned or are in advanced stages of planning. In particular, after implementation of the national RET in January 2010 with the mandate of generating 20 % or 45 TWh of electricity from RE sources by 2020, Australia has taken a wide range of initiatives to install large-scale wind energy plants around the country. In 2010, Australia's total installed capacity of wind energy was 1,880 MW, this being an increase of 167 MW from 2009. In the last decade, the growth rate of wind energy production was 30 % annually on average [20]. State-wise installed capacity of wind energy in Australia is given in Fig. 3.7 [21].

Large-scale generation of solar and wind energies reduces global warming as well as energy crisis worldwide and releases the pressure on other sources. Solar and wind energy generation is distributed over areas where solar radiation and wind speed are favourable to generate electricity. Australia has strong weather condition for solar PV and wind energy. Therefore, in large scale, such DER integration into the grid is imminent.

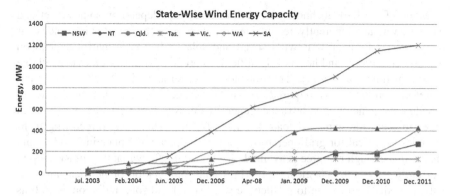

Fig. 3.7 State-wise installed wind energy capacity in Australia [21]

3.4 Distributed Energy Resources: Integration Challenges

DERs are electricity generation units, typically in the range of 3–50 kW, installed in low-voltage (below 25 kV) distribution systems at or near the end user. Solar and wind are the two promising sources of DER. They have the potential to improve reliability, PQ and global warming and reduce power generation, transmission and distribution costs. Grid-connected DERs support and strengthen the central bulk power station to meet the peak demands or to support major consumers. Moreover, DERs provide power to remote application where it is not possible to deliver power from traditional transmission and distribution lines [22, 23]. A typical snapshot of an integrated distributed energy system is shown in Fig. 3.8 [24].

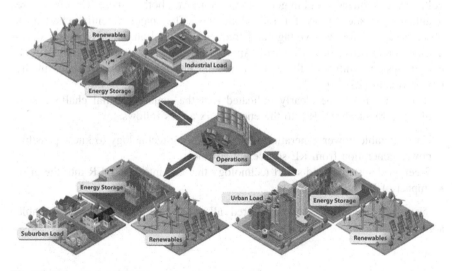

Fig. 3.8 Integrated distributed energy system [24]

Many of the DER technologies, such as those that depend on solar power and wind power, are inherently renewable in nature. Generation of wind and solar power varies frequently as wind speed and solar radiation primarily depend on weather conditions, and hence, it is difficult to predict the power outputs available at certain times of the day. The hour-to-day uncertainty in DEG is significant; therefore, prediction is required for load management and distribution or to schedule the generation in advance, generally hours to a full day ahead of time, in order to meet the expected load demand [22, 23]. Therefore, forecasting technologies is critical for grid operator to carry out operational planning studies and ensure that enough resource is available for managing the variability in RE output. Currently, physical, statistical and hybrid and combination of physical and statistical model is available to predict the wind speed and solar radiation as well as energy generation from these sources which can be used for adequate management of load demand systems [8].

The existing power systems are vulnerable for effective integration of DER. The lack of smart technologies in present network put the entire electricity system in risk. A modern smart power grid will bring benefits through seamless integration of DER to the smart grid. The use of smarter grid operations allows greater penetration of variable energy sources through more flexible management of the system. Fortunately, smart grid has the potential to mitigate some of the difficulties encountered by DER generation [8]. Therefore, it is a critical need today to introduce smart grid technology that accommodates RE to reduce overall GHGs, improves demand management, encourages energy efficiency, improves reliability and manages power more efficiently and effectively.

Large amount of intermittent generation from RE sources brings some uncertainties in both the generating sources and the DN behaviour. The variability nature in solar and wind energies has an impact on system operations, including voltage and frequency, and in general PQ. Moreover, both of these RE sources are unable to provide energy for the whole day. The major potential challenges observed are as follows: voltage and frequency regulation, reactive power compensation and active power control. Appropriate design of electrical circuits with control systems mitigates these problems and ensures PQ improvements in the power system [25, 26].

Therefore, it can be clearly indicated that the major potential challenges to facilitate large-scale DER into the energy mix are as follows:

- Unpredictable power generation; need forecasting technology to know possible power generation from RE sources in advance.
- Week grid system; need smart technology that can integrate DER into the grid.
- Impact of PQ and reliability.

Next sub-subsections have explored all the stated problems with suitable solutions.

3.4.1 Forecasting and Scheduling

The variability in DER requires knowing the relevant long-term weather patterns which can be used to develop better procedures and capabilities to facilitate integration into a "smart" national power grid. Accurate forecasting and scheduling systems are essential for appropriate and satisfactory use of DER and to establish sustainable load management systems for the smart grid.

The inherent mismatch between the DER output and the load may lead to significant energy wastage. For example, electric power can be generated from solar energy only at daytime, generally for a maximum of 8 h in a day, and fluctuates randomly with the movement of clouds. During daytime, load demand in residential areas is at its minimum which causes wastage of energy. A storage system is useful as it can store excess energy and provide power when energy shortages occur. The existing energy storing technologies include batteries, flywheels, supercapacitors and superconducting magnetic energy storage (SMES) [25, 26]. Integration of large-scale storage technology with the RE sources into the grid can ensure PQ and uniform power delivery. But the ancillary components (converters, filters, controllers, etc.) with the storage system have some effect on the overall power system.

Finally, a load demand management system is required that can be used by grid operators to accomplish operational planning and delivering smooth power supply to the consumers. A typical architecture of the load management system is shown in Fig. 3.9, which increases overall efficiency and quality of the system.

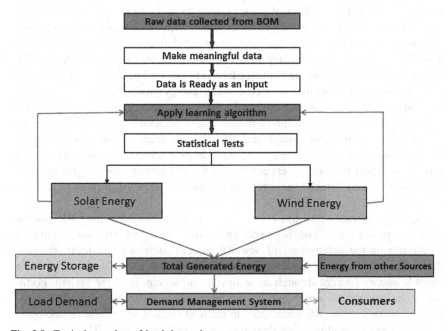

Fig. 3.9 Typical snapshot of load demand management system

Several research studies have been undertaken to assess solar irradiation and wind speed, as well as production of energy from these sources. Authors developed models [29] to forecast daily distribution of solar radiation and hourly distribution of wind speed as well as energy generation from these sources using ten popular regression algorithms. Meta-based learning random subspace (RSS), RegressionByDiscretization(RegDes), regression-based learning linear regression (LR), simple linear regression (SLR), statistical learning–based algorithm sequential minimal optimisation (SMO) regression (SMOReg), neural network–based multilayer perception (MLP), RBFNetwork (RBFN), lazy-based learning IBK, tree-based learning M5Rules and RepTree with bagging techniques [27, 28] were considered in the study for developing the prediction model. The most suitable algorithm was proposed based on the performance metrics [27] of the algorithms that include the correlation coefficient (CC), mean absolute error (MAE), root mean square error (RMSE) and computational complexity. Proposed algorithms with classical data-splitting options were used to predict the daily distribution of solar radiation and hourly distribution of wind speed. Details of the model are available in Ref. [29].

Initially, CC, MAE and RMSE have been measured for each of the models. From Fig. 3.10, it has seen that in terms of CC, the model developed with RSS performs the best. For RMSE measures, SMOReg is the best performing algorithm, while MLP is the worst. Therefore, it is really difficult to select the most suitable algorithm for this application. To select the most suitable model for this application, ranking performance for a given model has been estimated, and it has seen that RSS has ranked 1, while bagging technique has ranked 2 and MLP has ranked 10. Finally, the effect of ranking average and computational complexity was observed by changing the values of β that measures relative weighted performance. From Fig. 3.11, it has been observed that considering computational complexity and average accuracy RSS is the best performing algorithm for all β values and bagging technique with RepTree is the second performing algorithm to predict daily solar irradiation as well as solar energy production.

Similar models have been developed using the same ten regression algorithms with wind speed data. From the results, it was concluded that bagging technique with RepTree is the most suitable and RBFN is the second choice for predicting hourly distribution of wind speed as well as production of wind energy. The prediction model is therefore expected to play an important role for grid operator in controlling load demand management and smooth delivery of power supply to the consumers.

Existing electricity grid has experienced difficulties in integrating RE sources with the power grid. The high-voltage transmission grid imposes significant constraints on the deployment of new RE sources such as wind, solar and geothermal power. The use of smarter grid operations allows for greater penetration of variable energy sources through more flexible management of the system. Fortunately, an operational smart grid has the potential to mitigate some of the difficulties encountered by RE generation. In the next section, smart grid technologies have been explored along with their integration techniques with RE.

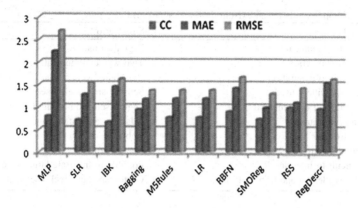

Fig. 3.10 Comparisons of performance metrics with different algorithms for prediction of daily solar irradiation

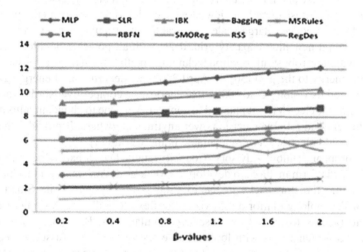

Fig. 3.11 Relative weighted performance of the algorithms with respect to β for prediction of solar irradiation

3.4.2 Smart Grid

Smart grid is the combination of centralised bulk power plants and distributed power generators that allows multidirectional power flow and information exchange. Its two-way power flow and communication systems create an automated and energy-efficient advanced energy delivery network. On the other hand, in traditional power systems, power flows only in one direction, that is, from generating station to customers via transmission and distribution networks [4, 30]. Brief comparisons between an existing grid and a smart grid are given in Table 3.1.

Table 3.1 Comparison between existing grid and smart grid

Existing grid	Smart grid
Mostly electromechanical	Digital in nature
One-way communication	Two-way communication
Mostly centralised generation	Distributed generation
Sensors are not widely used	Sensors are widely used
Lack of monitoring only manual	Digital self-monitoring
Failures and blackouts	Adaptive and intelligent
Lack of control	Robust control technology
Less energy efficient	Energy efficient
Usually difficult to integrate RE	Possible to integrate large scale RE
Customers have less scope to modify uses	Customers can check uses and modify

The European and North American vision for Smart Grid technology has evolved over the past decade and has begun to achieve a level of maturity. The USA's Centre for American Progress imparts a view of a clean electricity or clean energy "pipeline", which produces large-scale renewable electricity, delivers electricity nationwide on a new high-capacity grid, deals with all power generation and distribution with new robust information technology methods and allows consumers to contribute energy to the grid [30]. In April 2003, the department of energy (DOE), USA, declared its vision "Grid 2030" that energises a competitive North American Market place for electricity. It planned to connect everyone to abundant, affordable, clean, efficient and reliable electric power anytime, anywhere. It provides the best and most secure electric services available in the world [31].

The European Union Advisory Council established Smart Grid European Technology Platform in 2005, to develop the weak grid with smart grid technology and overcome the drawbacks in the existing power systems. The Smart Grid European Technology Platform [32] vision for flexible (fulfilling customer needs), accessible (access to all network users, particularly for RE sources and high-efficiency local generation with low carbon emission), reliable (assuring security and quality of supply) and economic (cost and energy efficient management) grid to meet the challenges and opportunities of the twenty-first century and fulfil the expectations of society.

Australia is lagging behind compared to the USA and Europe in an attempt to integrate RE sources and build smart grid infrastructure. The Australian government has already taken the initiative with large-scale investment to develop their electricity infrastructure as well as deploy smart grid technology to integrate large-scale RE into the grid [33].

The Australian government has adopted the Smart Grid, Smart City initiative and invested $100 million to create a large-scale smart grid platform, which optimises societal benefits by prioritising applications and undertaking a commercial-scale deployment that tests the business case and main technologies [11]. Integration studies are continuing to improve performances and facilitate large-scale RE penetration into the energy mix.

Fig. 3.12 Typical smart grid
model [34]

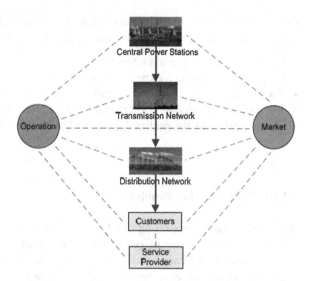

Smart grid technologies includes the following: automation technologies for the
smart power delivery; new advanced communication technologies; distributed
energy and storage technologies such as solar, wind turbine, fuel cell; advanced
metering infrastructure (AMI); power electronics-based controllers; appliances and
devices which are demand-response ready. Smart grid has the enhanced and robust
communication and computing capabilities that make this an attractive technology
for the future power system [34–36].

Recently, Zahedi [34] proposed a smart grid model that is useful to understand the
architecture and power flow of the smart grid as shown in Fig. 3.12. This model is
expected to be used as the basis in the design of smart grid infrastructure that defines
characteristics, requirements, interfaces and performance of the grid [34].

However, due to the intermittent nature of DER, integration of these sources
with the grid introduces technical challenges which need to be overcome to obtain
a sustainable, climate-friendly power system for the future. Issues that need to be
considered to integrate RE in particular wind and solar energy with the power grid
are standardised PQ, efficiency, reliability, cost of the energy conversion, appropriate load management, safety and security. Therefore, there is a prime need
today to reduce these potential technical challenges for a successful integration of
large-scale RE into the grid, though it is not an easy task. In the next section, the
observed potential challenges on large-scale RE integration have been presented.

3.4.3 Impacts of Renewable Energy into the Grid

Integration of large-scale DER in particular wind and solar energy with adequate PQ
into the grid is a challenging task due to the intermittent and weather-dependent

nature of these resources. The integration of variable generation sources presents unique challenges on system performance, and the key factors include [25]

- RE generator design parameters and power movers' type.
- RE power generation's expected types of run.
- Position of the RE plant's connection to the grid.
- Variability in production of RE sources with changing weather conditions.
- Characteristics of the grid including the loads connected to it.

3.4.3.1 Power Quality Problems

With the increased penetration of RE to the grid, the key potential technical challenges that effect quality of power observed include voltage fluctuation, power system transients and harmonics, reactive power and low power factor that detracts overall PQ of the power systems [25, 26] as shown in Fig. 3.13. These problems mostly occurred for wind and solar energy. Biomass, hydro- and geothermal energy sources are more predictable, and they have no significant problem on integration with the smart grid [25]. Details of the observed problems are given below:

Voltage fluctuation: Voltage fluctuation or instability as well as voltage sags/dips, noise, surges/spikes and power outages is the common problem encountered during integration of large-scale solar or wind energy into the grid. Variability in wind speed or solar irradiation with time is grid connection issues, and faults during operations and starting of large motors, etc., are also responsible. Large penetration of solar or wind power can lead to voltage control or the stability problem of power systems [16]. Periodic disturbances to the network voltage are denoted as flicker. The level of flicker is quantified by the short-term flicker severity value Pst, and allowable voltage change as a function of frequency is Pst = 1 [37, 38]. IEC Standard 61000-4-15 [39] provides a functional and design specification for flicker measuring instruments to measure the correct flicker perception level for all practical voltage fluctuation waveforms. The IEC Standard

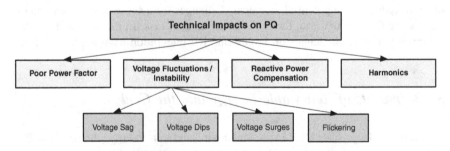

Fig. 3.13 Major potential technical impacts of integrating RE into the grid

61000-4-21 [40] provides a uniform methodology that ensures consistency and accuracy in the assessment of PQ characteristics of grid-connected wind turbines.

Reactive power compensation: The consumption of reactive power by induction generators is a common problem which affects the grid PQ. An induction generator requires an increasing amount of reactive power as the amount of power generated increases, and it is essential to provide reactive power locally as close as possible to the demand levels. Due to the fluctuations in the active and reactive power, the voltage at PCCs fluctuates. The most widely used reactive power compensation is capacitor compensation, which is static, low cost and readily available in different sizes [37, 41]. Reactive power compensation is typically implemented by using a fixed capacitor, a switched capacitor or a static compensator [42]. The power factor of the wind turbine can be improved significantly by appropriate compensation that enhances overall efficiency and voltage regulation of the system. Precise reactive power compensation considering proper size and proper control can remove voltage collapse and instability of the power system and enhances the overall operation of wind turbines.

Harmonic distortion: Power electronic devices, together with operation of nonlinear appliances, inject harmonics into the grid, which may potentially create voltage distortion problems. These results increase power system heat losses and reductions in the life of nearby connected equipment. Harmonic currents create problems both on the supply system and within the installation [43, 44]. Harmonic voltages cause voltage distortion, and zero-crossing noise in the network. The degree of distortion of an AC voltage or current is known as the total harmonic distortion (THD) and defined as the ratio of the sum of the squares of the individual harmonics to the fundamental harmonics as expressed by [43]:

$$\text{THD} = \frac{\sqrt{i_2^2 + i_3^2 + i_4^2 + \ldots\ldots + i_{n-1}^2}}{i_1} \tag{3.2}$$

Harmonics is one of the most dominant attributes that need to be kept to a minimum level to ensure good PQ of networks. Harmonic distortion can be minimised by good control algorithm design in the current control loop. Different types of filters are also used to mitigate harmonic distortion. The limits of different harmonic orders are specified in the Australian standard AS 4777 [45].

Appropriate design of electrical circuits with control systems mitigates voltage fluctuations, harmonic distortion, reactive power compensation and power factor improvements and ensures PQ improvements in the power system. Advanced inverter, controller and interconnection technology are starting to be used that allows RE to operate safely with the utility. Custom power devices such as static var compensators (SVCs), shunt active power filters (static synchronous compensators (STATCOMs)), series active power filters (dynamic voltage regulators (DVRs)) and a combination of series and shunt active power filters (unified power quality conditioners (UPQCs)) are the latest developments in interfacing devices between grids and consumer appliance. These devices reduce voltage/current disturbances and improve the PQ by compensating the reactive and harmonic power generated or absorbed by the load [46, 47]. Extensive planning, design and research need to be undertaken in different areas to mitigate the problems introduced by integrating large-scale RE into the grid.

In the next section, existing research that investigates the observed potential technical challenges with their mitigation techniques on integrating large-scale DER into the energy mix has been explored.

3.4.3.2 Existing Research on Integrating DER with the Grid

Wind and solar energy is the most promising DER which are free from GHGs and encourage interest worldwide. Wind generation is one of the fastest growing and cost-effective resources among the different RE sources. Small-scale photovoltaic technology is also cost-effective in providing electricity in rural or remote areas, in particular a country like in Australia.

A. Solar Energy Integration

The asymmetrical solar irradiation due to weather conditions, seasonal variation and geographical location produces voltage fluctuations in the output power of PV systems at the point of common connection (PCC) to the DN. Performances of PV modules also depend on load resistance, solar irradiance, cell temperature, cell shading and the crystalline structure. Under high PV saturation solar irradiance, temperature and shading can cause drastic swings in network operational performance in the event of cloud cover or low light situations, hence producing voltage and power fluctuations and decreased PQ.

This voltage fluctuation is exacerbated in single-phase PV systems as voltage increases both in phase and in neutral, hence causing phase imbalance in the network. However, PQ problems do not depend only on irradiation, but are also based on the overall performance of the solar PV system including PV modules, inverters, power electronic converters and connected consumer loads, etc. [25, 48].

Irregular power flows from PV systems and the use of power electronic converters inject harmonics into the power system network. This harmonic current flowing through the impedances of a DN causes voltage distortion. The fundamental requirements of an inverter are to limit the harmonic distortion and ensure a constant output voltage. Nonlinear loads connected to the DN also introduce harmonics. Many loads connected to the power system network require reactive power. However, a PV inverter is not able to fulfil this reactive power demand, hence causing low power factor in the network [48, 49]. Therefore, the most common impacts due to PV energy integration into the grid are voltage fluctuations, harmonics and poor power factor.

Suitable mitigation measures must be applied to the PV systems side to reduce voltage fluctuations and harmonics injection, and improve the power factor of the network for large PV systems into the grid. Significant researches are undertaken by various agencies throughout the world to investigate and mitigate the impacts on power system networks and to deploy large-scale PV into the grid [48, 49].

Asano et al. [50] analysed the impact of high penetration of PV on grid frequency regulation which responds to short-term irradiance or transients due to clouds. However, the break-even cost of PV is high unless PV penetration reaches 10 % or higher [50]. Therefore, PV integration needs to be increased and its impacts must be identified and mitigated [46, 50]. Recent studies by Ergon Energy (power utility operator in Queensland), and Chant et al. [51] have explored the issues involved in small-scale PV penetration in urban networks. It was found that increased penetration exhibited increased voltage rise on LV networks and increased harmonic distortion, and as a result, load rejection occurs.

A comprehensive study was carried out by Fekete et al. [49] that analysed the harmonic impacts on both winter and summer seasons with 10 kW PV penetration on the DN. Measured results provided basic guidelines that included the following: harmonic distortion of the PV current was low during high generation, and harmonic distortion was high in the period of low generation; odd current harmonics have significant impacts compared to even harmonics, and voltage harmonics had negligible impact on the DN.

In summary, experimental observations for PQ impacts of the integrated PV system are given in Table 3.2 [52]. However, the effects may become more prominent with the increase in PV penetration.

B. Wind Energy Integration

Integration of wind energy into the grid creates potential technical challenges that affect PQ of the systems due to the intermittent nature of wind energy. With the increased penetration of wind energy into the grid, the major PQ problems encountered in wind farms due to the design variations in wind turbines are [42, 53]

- Uncontrollable reactive power consumption and low power factor.
- Variations in wind speed cause power fluctuations on the grid.
- In a weak grid, power fluctuations cause severe voltage fluctuations as well as significant line losses.
- Injection of harmonics into the grid which may potentially create voltage distortion problems.

Potential technical difficulties occur not only due to the design of wind turbine types but also due to the intermittent nature of the wind source, electrical equipment and the grid connection characteristics and also due to grid quality issues [42]. The interactions between the wind turbine, the power network and the capacitor compensation are essential aspects of wind generation to optimise reactive power as well as active power, power factor and harmonic impacts [42, 53]. Developments in turbine technology have allowed harnessing more energy from the wind by improving the turbine height and increasing the swept area with larger blade sizes. Improved blade design has allowed the harvesting of very low and very high wind speeds and also increases the amount of power per swept area. Power electronic

Table 3.2 Power quality impacts of the PV clusters on distribution networks [52]

PQ concern	Observation	Consideration	Impact
Voltage variation	1 to 2 % increase at light load and high solar irradiation	Network configurations Number of feeders Voltage regulation method	May exceed the standard limit
Voltage unbalance	1–2 % variation due to uneven distribution of PV inverters on three phases and shading effect	Geographical and electrical distributions of PV installations in the area	Minor impact
THD voltage	5th, 7th and 11th harmonics slightly increase	Harmonic content of the grid voltage Series impedance of the grid	Normally below the standard limit
THD current	Harmonic distortion cloud increases at low solar generation	PV inverter topology Design of current control loop for the inverter Grid stiffness	May exceeds the standard limit; undesirable switch-off of PV inverter
Flicker	May occur at fast alternations of clouds and sunshine	Grid impedance	No noticeable impact

converters used in wind turbines are the main cause of harmonic current. With low levels of wind energy penetration, the overall effect on smart distribution system operations is limited, and if the penetration levels increase, more advanced control of the power system will be required to maintain system reliability. Table 3.3 details the potential challenges in integrating both small-scale wind energy and large-scale wind energy into the grid.

Theoretical aspects of the flicker algorithm, wind turbine characteristics and the generation of flicker during continuous and switching operations of wind turbines are evaluated in many researches [41, 54, 55]. From simulation results, it was shown that voltage fluctuations were widely affected by the grid strength and ratio of grid internal impedance. The impact of wind speed, turbulence intensity, grid voltage quality and the number of turbines operating in a grid-connected system are investigated and observed higher flicker levels at low wind speeds which exposed a large number of voltage dips. Flicker emission increases with the increase in wind turbulence intensity.

Rosas [16] investigated the impacts of wind power on the power system and observed that power converters can actively control the reactive power consumption which increased the voltage stability of the power system. From the literature [47, 56], it was found that DFIGs are the most efficient design for the regulation of reactive power and the adjustment of angular velocity to maximise the output power efficiency. These generators can also support the system during voltage sags, though this converter-based system injects harmonic distortion into the systems. However, the newly proposed Z-source inverter (ZSI) can mitigate the PQ problems for future DG systems connected to the grid [56].

Table 3.3 Potential challenges of integration of wind energy with the smart power grid [16]

Integration scale		Problems	Causes
Large scale	Small scale	Steady-state voltage rise	Wind speed variation
		Overcurrent	Peaks of wind speed
		Protection error action	Peaks of wind speed
		Flicker emission during continuous operation	Dynamic operation of wind turbines
		Flicker emission during switching operation	Switching/start-up operation of generators
		Voltage drop	In rush current due to switching operation of generators
		Harmonics	Power electronic converters
	Power system oscillations		Inability of the power system controllers to cope with the power variations from the wind farm and loads
	Voltage stability		Reactive power limitations and excessive reactive power demand from the power system

Characteristics of harmonics into a wind energy integrated power system were investigated with variety of configuration and operating condition [57]. PQ behaviours, in particular voltage sags and harmonics injection into the network, were investigated in [58] on integrating wind energy into LV and medium-voltage networks.

In order to enhance the terminal voltage quality, SVCs were used for reactive power compensation of wind power induction generators [59]. The use of STATCOMs with modified control strategies during normal and transient conditions has been addressed in [60] and [61], respectively. STATCOMs are superior compared to other flicker mitigation methods such as SVCs and series saturated reactors, STATCOMs being faster, smaller and having better performance at low-voltage conditions [62, 63].

STATCOM-based control mechanism is used to reduce the PQ problems as well as harmonics on integrating wind energy into the grid [64, 65]. Hybrid battery and supercapacitor energy storage systems are expected to play a major role in power smoothing, PQ improvement and LV ride through in a wind energy conversion system [17]. Kook et al. [65] developed a simulation model implemented using the power system simulator for engineering (PSS/E) that explored potential mitigation techniques to reduce the level of impacts on integrating wind energy into the grid through application of an energy storage system (ESS) [65]. Moreover, storage can play the vital role by load shifting to support the peak load demand when solar and wind are unable to generate energy.

Finally, the investigation (both experimental and simulated environments) was conducted to identify the impacts of large-scale RE integration into the grid, and suitable mitigation measures were proposed to reduce the level of impacts that ensure adequate PQ in the DN.

3.4.3.3 Impacts of DER Integration: A Case Study

Impacts of solar PV integration into the grid were investigated in experimental and simulation mode.

A. Experimental Analysis

To analyse the impacts of PV integration into the grid, experiments were undertaken at the renewable energy integration facility (REIF), CSIRO in Newcastle, Australia [66]. PQ parameters such as voltage fluctuations, reactive power compensation, harmonics and power factor of networks were investigated with varying PV penetration and load conditions. Details of the experiments were available in Ref. [67].

From the experimental results, it was observed that increased PV penetration causes voltage rise at the load and injects harmonics into the network which cause malfunctioning of devices and deteriorates the PQ of the network. The harmonic content of the network increases with the increase in PV penetration and system size. Figure 3.14 clearly indicates that neutral current is very large for both 11.3 and 7.5 kW PV penetrations and fluctuates significantly from the original sinusoidal wave. Experimental results showed that the neutral current for a PV system is large compared to an individual PV module and inject harmonics into the unbalanced system. It has also been observed that harmonics injection increases with increasing PV penetration, that is, with the increase in PV penetration from 7.5 to 11.3 kW.

Experimental results showed that the waveform of phase voltages is not purely sine waves and they also differ in amplitude and phase angle for both 7.5 and 11.3 kW PV integrated systems and observed voltage fluctuations as well as harmonics due to PV and loads.

From the FFT analysis, it was found that harmonic injections from all even harmonics are within the range stated in AS4777 standard [45]. However, the 3rd and 9th harmonics exceeded the regulatory standard, while injections from the 7th and 15th harmonics just reached the threshold levels as shown in Fig. 3.15. All other harmonics are within the allowable limits.

Minor voltage harmonics were observed in the network due to PV connections and load conditions. Observed THD is only 0.0128 (1.28 %), and second harmonics cause the most effect which is only 0.6 % of the fundamental voltage. It can be seen that in a few cases, the current harmonic distortion is beyond the allowable limits of AS 4777 [45].

B. Simulation Analysis

Considering reliability and flexibility of the modelling analysis and facilitate large-scale PV integration into the grid, different modelling case scenarios were developed using PSS Sincal [68] under the following configurations:

- Case 1: PV 9.5 kW, microturbine 28 kW, load 30 kW (considered harmonics injection only at PV).

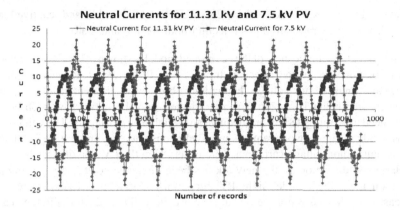

Fig. 3.14 Neutral currents with 11.3 and 7.5 kW PV penetrations

- Case 2: PV 19.1 kW, microturbine 28 kW, load 30 kW (considered harmonics injection only at PV).
- Case 3: PV 30.0 kW, microturbine 28 kW, load 30 kW (considered harmonics injection only at PV).
- Case 4: PV 30.0 kW, microturbine 28 kW, load 30 kW (considered harmonics injection both for PV and load).
- Case 5: PV 30.0 kW, microturbine 28 kW, load 60 kW (considered harmonics injection both for PV and load).
- Case 6: PV 30.0 kW, load 60 kW (considered harmonics injection both for PV and load).
- Case 7: PV 30.0 kW, microturbine 28 kW, active load 60 kW and reactive load 30 kVar (considered harmonics injection both for PV and load).

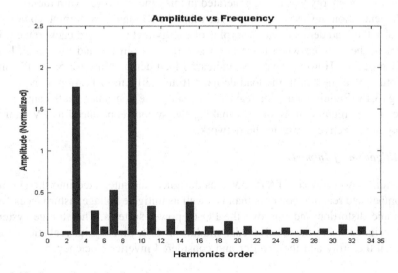

Fig. 3.15 Current harmonic distortion

- Case 8: PV 30.0 kW, microturbine 28 kW, active load 30 kW and reactive load 60 kVar (considered harmonics injection both for PV and load).

The typical schematic view of the developed model is shown in Fig. 3.16 in which major components are the infeeder, transformer, PV arrays, load, microturbine, busbar and lines. Minor voltage fluctuations observed in different case scenarios, however, significant harmonic currents injected into the network was observed which need to be mitigated for an adequate and reliable power supply. The voltage on the HV side of the network remains relatively constant even under high PV penetration, but minor voltage variations were observed in the LV side of the DN.

Harmonic distortion on the network has been demonstrated to have a minimal effect on the system under light PV penetration. Modelling results show that increases in PV generation cause an increase in THD. THD for different case scenarios is shown in Fig. 3.17. It can be concluded that THD increases with the increase in PV generation into the network and THDs for Case 1, Case 2 and Case 3 are, respectively, 1.1, 2.2 and 3.8 % across the LV network and 1.3, 2.6 and 4.4 % across the point of the PV inverter connection. It has been observed that harmonic injection from the PV in Case 3 almost reaches the threshold limit of 5 % defined by IEC 61000-2-4 as well as Australian standard AS 4777 [45]. Load harmonics were added from Case 4 to Case 8 of the modelling to analyse the adverse harmonic impacts due to connected consumer loads. Figure 3.17 indicates that THD increases significantly due to added load harmonics. The only difference between Case 4 and Case 5 is the amount of load connected, and from Fig. 3.17, it is clearly indicated that harmonic distortion increases with the increase in load.

Grid supplied all the required reactive power as the PV and the microturbine could not supply any reactive power in the network for Case 1. The PV system and the microturbine can fulfil the load demand only from 11:30 a.m to 1:30 p.m as maximum PV energy was being generated in this time. However, with the increase in PV generation in Case 3, it can be seen that the PV and microturbine combined can fulfil the load demand with a surplus of energy that can be fed back to the grid except in the evening when there is no generation from PV and maximum load demand occurs. However, with the increased load demand in Case 5, the PV and microturbine cannot fulfil the load demand from 6:30 p.m to 11:45 p.m as shown in Fig. 3.18. Requirements for reactive power by the consumer loads cause poor power factor regulation as well as making the system unbalanced as PV cannot supply any reactive power to the network.

C. Mitigation of Impacts

A custom power device STATCOM was designed and integrated into the system to compensate reactive power demand as well as mitigate voltage disturbances and harmonic distortion and improve the PQ of power systems. The storage system was designed using the DC-Infeeder module in PSS Sincal with realistic losses between the array and the inverter and considered inverter efficiency.

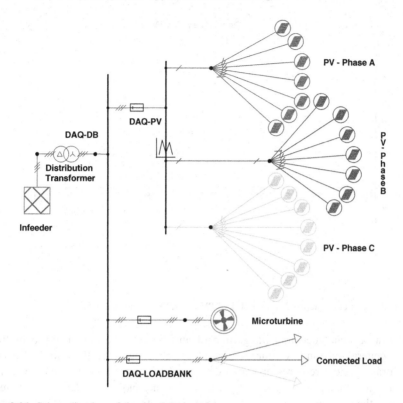

Fig. 3.16 Schematic view of the developed model

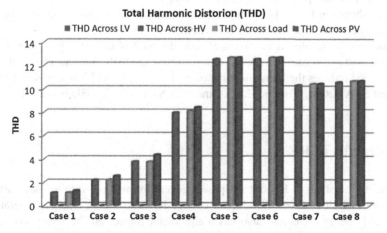

Fig. 3.17 Total harmonic distortion (THD)

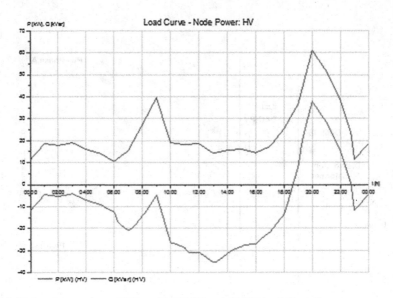

Fig. 3.18 Load curve across DT in a 30-kW PV and 60-kW load (Case 5)

It can be concluded that integration of an ESS into the model improves the power generation from PV sources and reduces dependency on the grid. However, the storage system requires power to charge itself during the day while PV generation was maximum with minimum load demand and it can deliver its stored power in the evening and reduces grid dependency. From simulation results, it can be seen that energy generation was increased after integrating energy storage. Integration of an optimised STATCOM compensates reactive power of the network as shown in Fig. 3.19, which improves the power factor as well as reduces voltage fluctuation and harmonic distortion.

From model results, it can be seen that the STATCOM stabilised the phase voltages, and in this case, the phase voltages were close to the nominal rated voltage. STATCOM reduces the harmonic distortion significantly, and it was observed that without STATCOM, THD is 12.5 %, and with STATCOM, THD is 0.2 %.

3.5 Benefits of RE

Increasing the share of RE in total power generation will not only reduce carbon emission and slow down the climate change but also reduce energy generation costs and energy crisis and also has a significant impact on the socio-economic development worldwide.

Authors carried out a feasibility analyses using the hybrid optimisation model for electric renewable (HOMER) [69] which showed the potential of RE in Australian region considering the production cost, cost of energy and emission

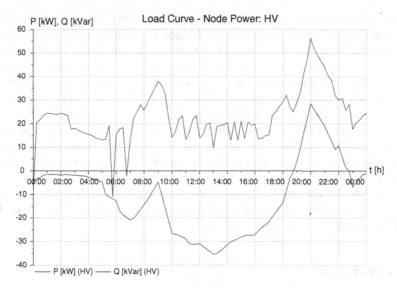

Fig. 3.19 Load curve across DT in a 30-kW PV and 60-kW load with storage and STATCOM (Case 5c)

reduction [70]. From the study, it is clearly observed that Australia has enormous potential for substantially increased use of RE. A large penetration of RE sources into the national power system would reduce CO_2 emissions significantly, contributing to the reduction in global warming. It is also indicated in various studies that RE not only reduces the GHG emission but also plays a major role in reducing energy crisis worldwide as these sources are unlimited and reduces costs of energy generation [70, 71].

From the statistical analysis [70], it was estimated that for wind energy generation, Tasmania is ranked 1 (most suitable) and the Northern Territory is ranked 7 (least suitable) out of the seven states of Australia. On the other hand, it can be seen that the Northern Territory is ranked 1 and Tasmania is ranked 7 out of the seven states of Australia for solar energy generation. The three best potential locations for wind energy generation in Australia are Macquarie Island in Tasmania, Wilsons Promontory in Victoria and Cape Leewin in Western Australia. The three best potential locations for solar energy generation are Weipa in Queensland, Alice Springs in the Northern Territory and Karratha in Western Australia. From the study, it was also evident that Flinders Reef in QLD is the best place considering the combined energy generation from both solar and wind resources; contribution from RE is 90 %. This proposed model will be of benefit to researchers and power utilities to further assess the prospects of RE sources and suitable locations for both wind and solar energy generation in Australia and thus assist to achieve the national goal of introducing 20–25 % energy from RE sources by 2020 [70, 71].

RE generation would also bring indirect benefits like income generation, employment creation and improvements in local air quality and other enhancements for quality of life. According to Clean Energy Australia 2010 report by Clean Energy Council, more than 55,000 jobs are expected to be created in RE by 2020. Due to increased growth of PV, Germany has employed 40,200 in 2006 and 50,700 in 2007 in photovoltaic sector. Job creation is an important part of economic development activity and strong economies as the creation of employment not only benefits the community through the income earned from those jobs, and it generates spin-off benefits known as the multiplier effect [5]. A review of some 30 studies of employment in the energy sectors in North America showed that RE projects can create twice as many jobs as conventional energy projects, per dollar invested [72]. According to [73], the economic advantages of RE technologies are twofold: (1) they are labour intensive, so they generally create more employments for the same amount of investment than conventional electricity generation technologies and (2) they use primarily native resources, so most of the energy per dollars can be kept at home.

3.6 Conclusions

Recent environmental awareness resulting from the conventional power station has encouraged interest in the deployment of large-scale RE in the energy mix. Current power systems are not capable of mixing RE sources as the systems were not developed for such integration. The use of smart grid operations allows for greater penetration of variable energy sources through more flexible management of the system. RE sources not only reduce GHG emission significantly which plays a key role in developing a sustainable climate-friendly environment but also reduce the energy generation costs and energy crisis worldwide.

Therefore, considering the current scenario, substantial research, planning and development work have undertaken worldwide, to facilitate large-scale integration of RE into the energy mix. Large penetration of RE sources into the grid causes significant voltage and power fluctuation, harmonics injection as well as frequency deviation in the network which reduces the PQ. Appropriate design and control of power electronic devices not only ensures the reliability and availability of power delivery but also improves the voltage stability and power system stability and thus ensures continuous increase in RE into the power network. Findings of this study are expected to be used as guidelines by the policy makers, manufacturers, industrialists and utilities for deployment of large-scale RE into the energy mix.

References

1. IEA statistics: CO_2 emissions from fuel combustion, International Energy Agency (IEA) (2011)
2. An Atlas of Pollution (2011), US Energy Information Administration, Feb 2011. Online available: http://sustainabletransition.blogspot.com.au/2011/02/atlas-of-pollution.html, access date 10 July 2012

3. Energy in Australia 2010, Technical report, Department of resources, energy and tourism, Government of Australia, 2010. Online available: http://adl.brs.gov.au/data/warehouse/pe_abarebrs99014444/energyAUS2010.pdf, access date 10 July 2012
4. Mitchell KO (2005) Optimisation of the applications of sustainable energy systems. PhD Thesis, College of Science, Technology and Environment, University of Western Sydney, Australia
5. Akella AK, Saini RP, Sharma MP (2009) Social, economical and environmental impacts of renewable energy systems. Proc Renew Energy 34:390–396 ELSEVIER
6. Gol O (2008) Renewable energy: panacea for climate change? In: Proceedings of ICREPQ'08, Santander, Spain, March 2008
7. Zabihian F, Fung A (2009) Fuel and GHG emission reduction potentials by fuel switching and technology improvement in the Iranian electricity generation sector. Int J Eng 3:159–173
8. Grid integration of large-capacity renewable energy sources and use of large-capacity electrical energy storage. Technical report, International Electrotechnical Commission (IEC), Oct 2012
9. REN21 (2011): Renewables 2011 Global status report
10. Clean energy initiative overview. Technical report, Department of Resources, Energy and Tourism, Australian Government.
11. Smart grid, smart city: a new direction for a new energy era. Technical report: Department of the Environment, Water, Heritage and the Arts, Australia, 2009
12. iGrid intelligent grid, Technical report. Available at: http://www.igrid.net.au/
13. Jenkins N, Allan R, Crossley P, Kirschen D, Starbac G (1995) Embedded generation, The Institution of Electrical Engineers, London, ISBN: 0-85296-774-8, 978-0-85296-774-4
14. Nielsen R Solar radiation, Technical report. Online available: http://home.iprimus.com.au/nielsens/solrad.html
15. PV in Australia 2011, Technical report, Australian PV Association, May 2012
16. Rosas P (2003) Dynamic influences of wind power on the power system. PhD Thesis, Orsted Institute, Electric Power Engineering, Technical University of Denmark, Denmark, Mar 2003
17. Chowdhury MM, Haque ME, Aktarujjaman M, Negnevitsky M, Gargoom A (2011) Grid integration impacts and energy storage systems for wind energy applications: a review. In: Proceedings of the IEEE power and energy society general meeting, USA, 24–29 July 2011
18. Kamau JN, Kinyua R, Gathua JK (2010) 6 years of wind data for Marsabit, Kenya average over 14 m/s at 100 m hub height: an analysis of the wind energy potential. J Renew Energ 35:1298–1302 ELSEVIER
19. Fox B, Flynn D, Bryans L, Jenkins N, Milborrow D, O'Malley M, Watson R, Anaya-Lara O (2007) Wind power integration: connection and system operational aspects. The Institution of Engineering and Technology (IET), London, ISBN 978-0-86341-449-7, 2007
20. Global wind report: the global status of wind power in 2011, Technical report: global wind energy council, 2011
21. Shafiullah GM, Oo MT Amanullah, S ABM Ali, Wolfs P (2013) Potential challenges of integrating large-scale wind energy into the power grid: a review. Journal of Rene and Sustainable Energy Reviews, Elsevier, vol. 20, pp 306–321
22. Capehart BL (2012) Distributed energy resources (DER), College of Engineering, University of Florida, [Online Available]: http://www.wbdg.org/resources/der.php, access date 5 Dec 2012
23. Chowdhury BH, Tseng C (2007) Distributed energy resources: issues and challenges, Editorial J Energ Eng, Sep 2007
24. Integrated distributed energy resource solutions, Technical report BPL Africa. Online available: http://www.bplafrica.com/show.aspx?id=98, access date 5 Dec 2012
25. Ming Z, Lixin H, Fam Y, Danwei J (2010) Research of the problems of renewable energy orderly combined to the grid in smart grid. In: Proceedings of the power and energy engineering conference (APPEEC 2010), Chengdu, China, 28–31 Mar 2010

26. Liserre M, Sauter T, Hung JY (2010) Future energy systems: integrating renewable energy sources into the smart power grid through industrial electronics. IEEE Ind Electron Mag 4(1):18–37
27. Witten IH, Frank E (2000) Data mining: practical machine learning tool and technique with java implementation. Morgan Kaufmann, San Francisco
28. Weka 3 (2009) The University of Waikato, New Zealand, Technical report, Available at: http://www.cs.waikato.ac.nz/ml/weka/ as at 15 Mar 2009
29. Shafiullah GM, Oo MT Amanullah, Jarvis D, Ali S, Wolfs P (2010) Potential challenges: integrating renewable energy with the smart grid. In: Proceedings of Australasian universities power engineering conference, (AUPEC 2010), Christchurch, New Zealand, Dec 2010
30. Hendricks B (2009) Wired for progress: building a national clean-energy smart grid. Technical report: Center for American Progress, USA, Feb 2009
31. Grid 2030 (2003) A national vision for electricity's second 100 Years. U.S. Department of Energy, Office of Electric Transmission and Distribution, July 2003, Available at: http://www.climatevision.gov/sectors/electricpower/pdfs/electric_vision.pdf
32. Smart Grids: European Technology Platform. Technical report, Available at: http://www.smartgrids.eu/documents/vision.pdf
33. Davidson M (2010) Smart grid Australia: an overview. Technical report, Wessex Consult, Australia, Feb 2010
34. Zahedi A (2011) Developing a system model for future smart grid. In: Proceedings in 2011 IEEE PES innovative smart grid technologies conference, ISGT Asia 2011, Perth, Australia, 13–16 Nov 2011
35. Fang X, Misra S, Xue G, Yang D (2011) Smart grid—the new and improved power grid: a survey. Int J IEEE Commun Surv Tutorials 99:1–37
36. Gungor VC, Sahin D, Kocak T, Buccella C, Cecati C, Hancke GP (2011) Smart grid technologies: communication technologies and standards. IEEE Trans Ind Inform 7(4):529–538
37. Linh NT (2009) Power quality investigation of grid connected wind turbines. In: Proceedings of the 4th IEEE conference on industrial electronics and applications, China, 25–27 May 2009
38. Bossanyi E, Saad-Saoud Z, Jenkins N (1998) Prediction of flicker produced by wind turbines. J Wind Energy 1:35–51
39. IEC Standard 61000-4-15 (2003) Electromagnetic compatibility (EMC)—part 4: testing and measurement techniques—section 15: Flickermeter—Functional and design specifications, International Electrotechnical Commission, 2003
40. IEC Standard 61000-4-21 (2001) Wind turbine generator systems—part 21: measurement and assessment of power quality characteristics of grid connected wind turbines, International Electrotechnical Commission, 2001
41. Ei-Tamaly HH, Wahab MAA, Kasem AH (2007) Simulation of directly grid-connected wind turbines for voltage fluctuation evaluation. Int J Appl Eng Res 2(1):15–30
42. Muljadi E, Butterfield CP, Yinger R, Romanowitz H (2004) Energy storage and reactive power compensator in a large wind farm. In: Proceedings of 42nd AIAA aerospace science meeting and exhibit, Reno, Nevada, 5–8 Jan 2004
43. Harmonics: causes and effects, power quality application guide, Copper Development Association. Technical report. Online available: http://www.leonardo-energy.org/webfm_send/115
44. Lewis SJ (2011) Analysis and management of the impacts of a high penetration of photovoltaic systems in an electricity distribution network. In: Proceedings of the innovative smart grid technologies conference (ISGT 2011), Perth, Australia, 13–16 Nov 2011
45. Standards Australia, AS 4777 (2005) Grid connection of energy systems via inverter, online available: http://www.saiglobal.com
46. Eltawil MA, Zhao Z (2010) Grid-connected photovoltaic power systems: technical and potential problems—a review. Int J Renew Sustain Energy Rev 14:112–129 Elsevier

47. Khadem SK, Basu M, Conlon MF (2010) Power quality in grid connected renewable energy systems: role of custom power devices. In: Proceedings of the international conference on renewable energies and power quality (ICREPQ'10), Granada, Spain, 23–35 Mar 2010
48. Albarracin R, Amaris H (2009) Power quality in distribution power networks with photovoltaic energy sources. In: Proceedings of the 8th international conference on environment and electrical engineering, IEEE, Karpacz, Poland, 10–13 May 2009, online available: http://eeeic.eu/proc/papers/131.pdf
49. Fekete K, Klaic Z, Majdandzic L (2012) Expansion of the residential photovoltaic systems and its harmonic impact on the distribution grid. Int J Renew Energy 43:140–148 Elsevier
50. Asano H, Yajima K, Kaya Y (1996) Influence of photovoltaic power generation on required capacity for load frequency control. IEEE Trans Energy Convers 11:188–193
51. Chant T, Shafiullah GM, Oo MT Amanullah, Harvey B (2011) Impacts of increased photovoltaic panel utilization on utility grid operations: a case study for central Queensland. In: Proceedings of the innovative smart grid technologies conference (ISGT 2011), Perth, Australia, 13–16 Nov 2011
52. Katirael F, Mauch K, Dignard-Bailey L (2007) Integration of photovoltaic power systems in high-penetration clusters for distribution networks and mini-grids. Int J Distrib Energy Resour 3(3):207–223
53. Muljadi E, Butterfield CP (2005) Self excitation and harmonics in wind power generation. In: Proceedings of 43rd AIAA aerospace science meeting and exhibit, Reno, Nevada, 10–13 Jan 2005
54. Thiringer T, Petru T, Stefan L (2004) Flicker contribution from wind turbine installations. IEEE Trans Energy Convers 19:157–163
55. Larsson Ake (2002) Flicker emission of wind turbines during continuous operation. IEEE Trans Energy Convers 17:114–118
56. Dehghan SM, Mohamadian M, Varjani AY (2009) A new variable speed wind energy conversion system using permanent magnet synchronous generator and Z-source inverter. IEEE Trans Energy Convers 24:714–724
57. Papathanassiou SA, Papadopoulos MP (2006) Harmonic analysis in a power system with wind generation. IEEE Trans Power Deliv 21:2006–2016
58. Vilchez E, Stenzel J Wind energy integration into 380 kV system—impact on power quality of MV and LV networks. In: Proceedings of international conference on renewable energies and power quality (ICREPQ)
59. Ra Jambal K, Umamaheswari B, Chellamuthu C (2005) Steady state analysis of grid-connected fixed-speed wind turbines. Int J Power Energy Syst 25:230–236
60. Chen WL, Hsu YT (2006) Controller design for an induction generator driven by a variable-speed wind Turbine. IEEE Trans Energy Convers 21:625–635
61. Gaztanaga H, Etxeberria-Otadui I, Ocnasu D, Bacha S (2007) Real time analysis of the transient response improvement of fixed-speed wind farms by using a reduced-scale STATCOM prototype. IEEE Trans Power Syst 22:658–666
62. Sannino A, Svensson J, Larsson T (2003) Power-electronic solutions to power quality problems. Electric Power Syst Res 66:71–82 Elsevier
63. Larsson T Voltage source converters for mitigation of flicker caused by arc furnaces. PhD thesis
64. Yuvaraj V, Deepa SN, Rozario APR, Kumar M (2011) Improving grid power quality with FACTS device on integration of wind energy system. In: Proceedings of 5th Asia modelling symposium (AMS), pp 157–162, Kuala Lumpur, Malaysia, 24–26 May 2011
65. Kook KS, Liu Y, Atcitty S (2006) Mitigation of the wind generation integration related power quality issues by energy storage. J Electric Power Qual Utilisation 12:77–82
66. Renewable Energy Integration Facility (2011) CSIRO, Technical report, Oct 2011. Online available: http://www.csiro.au/Outcomes/Energy/Renewable-Energy-Integration-Facility.aspx
67. Shafiullah GM, Oo MT Amanullah, Ali S, Azad S, Arif M, Moore T (2012) Experimental analysis of harmonics on utility grid with PV penetration. In: Proceedings of the international conference on electrical and computer systems (ICECS 2012), Ottawa, CA, USA, 22–24 Aug 2012

68. PSS Sincal (2012) PSS Product Suite, Siemens. Online available: http://www.energy. siemens.com/us/en/services/power-transmission-distribution/power-technologies-international/software-solutions/pss-sincal.htm
69. HOMER—analysis of micro power system options. Online available at: https://analysis. nrel.gov/homer/
70. Shafiullah GM, Oo MT Amanullah, Ali S, Wolfs P (2012) Prospects of renewable energy: a feasibility study in the Australian context. Int J Renew Energy 39:183–197 Elsevier
71. Shafiullah G, Oo A, Jarvis D, Ali ABMS, Wolfs P (2010) Economic analysis of hybrid renewable model for subtropical climate. Int J Therm Environ Eng (IJTEE) 1:57–65
72. Campbell B, Pape A (2009) Economic development from renewable energy. Discussion Paper, Pembina Institute for Appropriate Development
73. Dollars from sense: the economic benefits of renewable energy, National Renewable Energy Laboratory. Online available: http://www.nrel.gov/docs/legosti/fy97/20505.pdf

Chapter 4
Energy Storage: Applications and Advantages

Mohammad Taufiqul Arif, Amanullah M. T. Oo
and A. B. M. Shawkat Ali

Abstract Energy storage (ES) is a form of media that store some form of energy to be used at a later time. In traditional power system, ES play a relatively minor role, but as the intermittent renewable energy (RE) resources or distributed generators and advanced technologies integrate into the power grid, storage becomes the key enabler of low-carbon, smart power systems for the future. Most RE sources cannot provide steady energy supply and introduce a potential unbalance in energy supply and load demand. ES can buffer sizable portion of energy generated by different intermittent RE sources during low demand time and export it back into the network as required. ES can be utilized in load shifting, energy management and network voltage regulations. It can play a large role in supplementing peaking generation to meet short-period peak load demand. ES technologies are classified considering energy and power density, response time, cost, lifetime and efficiency. Different application requires different types of ES system (ESS). IEEE 1547 and AS 4777 provide guideline to connect RE and storage into the distribution network. Based on the standards, utility operators plan in gradual integration of RE into the grid. Storage can play significant role in reduction in greenhouse gas (GHG) emission by maximizing RE utilization. As the utility operator needs to support costly peak load demand which could be supported by storage and as a consequence, storage can help in energy cost reduction. Although, the present cost of storage considered a barrier for extensive use, however, research is going on for low-cost, high-performance storage system. Therefore, in the low-carbon future power system, ES will play a significant role in increasing

M. T. Arif (✉) · A. M. T. Oo · A. B. M. S. Ali
School of Engineering and Technology, Central Queensland University, Bruce Highway,
Rockhampton, QLD 4702, Australia
e-mail: m.arif@cqu.edu.au

A. M. T. Oo
e-mail: a.maungthanoo@cqu.edu.au

A. B. M. S. Ali
e-mail: S.Ali@cqu.edu.au

A. B. M. S. Ali (ed.), *Smart Grids*, Green Energy and Technology,
DOI: 10.1007/978-1-4471-5210-1_4, © Springer-Verlag London 2013

grid reliability and enabling smart grid capabilities for sustainable future by balancing RE output.

This chapter explained various energy storage (ES) technologies, their applications, advantages, cost comparison and described integration of storage into the grid. Two case studies are explained in this chapter to illustrate the advantages of ES. First one explained storage advantage in distribution transformer (DT) utilization and fluctuation minimization. Other one explained economical and environmental benefit of ES. Lastly, future direction of ES system (ESS) also explained.

4.1 Introduction

Demand of ESS increased due to the technological development to use with intermittent RE and to support the growing use in electric vehicle (EV). ES can augment generation from renewable distributed energy generators (DEG) such as solar and wind in three ways. Firstly, it can be used for stabilizing purposes by enabling DEGs to run in the acceptable limit and minimizes energy fluctuations. Secondly, proper-sized ES can ride through periods by load shifting when DEGs are unable to generate energy. Thirdly, ES can permit non-dispatchable DEG to operate as a dispatchable unit by supporting timely load demand in network. Large storage lets RE producers to store surplus energy and supply to the grid when load demand goes high and also balances the demand and supply.

In Australia, approximately 10 % of Queensland's electricity network has been built to support only the extreme peak loads [1], similarly other utility operators maintain costly short time generators to support peak load. By integrating proper-sized ES with RE, this peak load demand can be minimized and eventually helps to reduce the cost of energy (COE).

In Queensland, Australia, peak demand generally occurs between 4:00 PM and 8:00 PM, when most householders return home and turn on energy-intensive appliances [1]. Queensland's electricity demand will continue to grow more than 3.5 % per year [1]. Utility operators need to maintain additional facility to support the peak demand. Also during peak demand, RE such as solar and partly wind is not able to generate energy. Therefore, storage is the key enabler to support future load demand especially in peak demand period and to extend the use of RE. There are different ES technologies available and suitable for different applications.

4.2 Different Energy Storage Technologies

Different ES technologies coexist and different characteristics make them suitable for different applications. ES is now seen more as a tool to improve power quality in power systems, assist in power transfer and to enhance system stability. Recent developments and advances in ES and power electronics technology make ES applications a feasible solution for modern power applications. In an AC (Alternating current) system, electrical energy cannot be stored electrically; however, energy can be stored by converting and storing it electrochemically, electromagnetically, kinetically or as potential energy. Each ES technology contains a power conversion unit. Two factors characterize the application of ES technology. One is the amount of energy that can be stored and other is the rate of energy transfer to/from the storage devices.

ES technologies can be classified considering energy and power density, response time, cost, lifetime, efficiency or operating constraints. Among different forms of ESS, pumped hydroelectric storage, compressed air energy storage, thermal energy storage, flywheel, hydrogen, different types of batteries, capacitors, superconducting magnetic energy storage are suitable for different types of applications. Different ESSs are explained below.

4.2.1 Battery Energy Storage System (BESS)

Battery is one of the most cost-effective ES technologies available today where energy stored electrochemically [2]. BESS is a modular technology and one of the promising storage technologies for power applications such as regulations, protection, spinning reserve and power factor correction [3]. Batteries are charged/stored energy by internal chemical reaction when potential is applied to the terminal and discharge energy by reverse chemical reaction. Battery stores DC charge, and therefore, converter is required to interface with the AC system.

There are a number of battery technologies under consideration for large-scale application. Lead-acid batteries is an established and mature technology that can be designed for bulk ES or for rapid charge or discharge application. Other battery technologies are nickel-metal hydride (Ni-MH), nickel–cadmium (Ni–Cd), lithium-ion (Li-ion), sodium–sulphur (NaS) and flow battery (FBs). There are three types of FBs: vanadium redox battery (VRB), polysulphide bromide battery (PSB) and zinc bromide battery (ZnBr). Figure 4.1 shows different battery systems.

Lead-acid   Lithium-ion   Vanadium-Redox

Fig. 4.1 Different battery technology. *Source* http://electrical-engineering-portal.com/wp-content/uploads/lead-acid-battery-construction.jpg, http://www.okta.net/_okta/okta_board_view.asp?idx=1612&flag=world_talk, http://peswiki.com/images/thumb/7/75/VanadiumRedoxBattery.png/333px-VanadiumRedoxBattery.png

BESS can response very fast and cost is comparatively low. Lead-acid battery is suitable for backup power application. Ni–Cd is suitable for peak shaving application and to support during voltage sag. NaS battery has more energy density therefore has longer life and higher round-trip energy efficiency. NaS battery is suitable for energy management and power quality. VRB has fast response and it can be used for load levelling, peak shaving and integrating RE resources. PSB can be used in load levelling, peak shaving and integrating RE resources. PSB batteries particularly useful for frequency response and voltage control. ZnBr batteries are suitable for smoothing out fluctuations, and load management.

Lead-acid battery is sensitive in operating temperature, and best operating temperature is about 27 °C, has depth of discharge (DoD) limit and charge/discharge cycle limit. NaS needs to keep in elevated temperature between 300–350 °C [4]. VRB has lower power density.

The largest lead-acid battery installed in California has a capacity of 10 MW/40 MWh [5]. Largest NaS battery has a rating of 9.6 MW/64 MWh, and largest VRB has a rating of 1.5 MW/1.5 MWh [5].

4.2.2 Superconducting Magnetic Energy Storage (SMES)

It is a device that stores energy in the magnetic field generated by the DC (Direct current) current flowing through a superconducting coil. The inductively stored energy (E in joules) and the rated power (P in watts) are the common specifications of SMES and can be expressed by Eq. (4.1) [2]:

$$E = \frac{1}{2}LI^2 \quad P = \frac{dE}{dt} = \quad LI\frac{dI}{dt} = VI \tag{4.1}$$

where L is the inductance of the coil, I is the DC current flowing through the coil and V is the voltage across the coil. Energy can be drawn from SMES almost as an instantaneous response and can be stored or delivered over periods ranging from a fraction of a second to several hours.

SMES attracted attention due to their fast response and high efficiency (a charge–discharge efficiency over 95 %). Possible applications of SMES include load levelling, voltage stability, dynamic stability, transient stability, frequency regulation, transmission capacity enhancement and power quality improvement [2]. SMES system still costly compared to other ES technologies. It is sensitive to temperature and can become unstable in temperature change.

4.2.3 Super Capacitors Energy Storage (SCES)

Capacitors store accumulated positive or negative electric charges on parallel plates separated by dielectric materials. Capacitance (C) represented by the relationship between stored charge (q) and voltage between plates (V). Capacitance depends on the area of the plates (A), distance between plates (d) and permittivity of the dielectric (ε) as shown in Eq. (4.2) [2].

$$q = CV \quad C = \frac{\varepsilon A}{d} \quad E = \frac{1}{2}CV^2 \tag{4.2}$$

The amount of energy can be increased by increasing capacitance or voltage between the plates. However, voltage depends on the withstand strength of the dielectric, also impacted by the distance between plates. The total voltage change when charging or discharging capacitors is shown in Eq. (4.3) [2] where C_{tot} and R_{tot} are the total capacitance and resistance from a combined series/parallel configuration of capacitor cells to increase total capacitance and total voltage level.

$$dV = i * \frac{dt}{C_{tot}} + i * R_{tot} \tag{4.3}$$

Capacitors are used in many AC or DC applications. Capacitors are often used as very short-term storage with power converters. Capacitance can be added to the DC bus of motor drives or consumer electronics to provide additional capability to operate during voltage sags and momentary interruptions. DC capacitors are used as large-scale ES on distribution dynamic voltage restorer (DVR) that compensates for temporary voltage sags on the power distribution systems [6]. The disadvantage of capacitor is its low energy density.

Ceramic hyper-capacitors have both fairly high voltage withstand capacity (about 1 kV) and high dielectric strength. Ultra-capacitors are double-layer capacitors that have increased storage capacity and suitable for high peak power, low energy applications. Electrochemical double-layer capacitors (EDLCs) work similar as conventional capacitors but have very high capacitance ratings, long life cycle and better efficiency.

4.2.4 Flywheel Energy Storage (FES)

FES stores energy in a rotatory mass. Flywheel can be used to store energy for power systems when it coupled to an electric machine such as synchronous generator. Stored energy (E) depends on the moment of inertia (J) of the rotor and the square of the rotational velocity (ω) of the flywheel. Moment of inertia depends on the radius (r), mass (m) and length/height (h) of the rotor as shown in Eqs. (4.4) and (4.5) [2]. Figure 4.2 shows a FES application scheme.

$$E = \frac{1}{2}J\omega^2 \tag{4.4}$$

$$J = \frac{r^2 mh}{2} \tag{4.5}$$

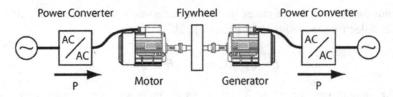

Fig. 4.2 General scheme of flywheel with two machines [7]

FES systems are able to provide very high peak power, high power and energy density and virtually have infinite number of charge–discharge cycles [7]. Flywheel has been considered for numerous power system applications, including power quality, peak shaving and stability enhancement applications and also for transportation applications. It requires cooling and there is power loss during ideal time.

4.2.5 Thermal Energy Storage (TES)

It involves storing energy in a thermal reservoir to use at a later time. TES system suitable for solar thermal power plants consists of synthetic oil or molten salt as heat ES. Heat collected from concentrated solar power plant (CSP) as shown in Fig. 4.3. CSP with TES can store thermal energy for period up to 15 h, thus improve flexibility of the grid and facilitate towards greater penetration of solar energy into the grid. TES can be used to increase reliability of intermittent RE sources.

Other types of TES utilize electricity during off-peak periods and stores energy as hot or cold storage in underground aquifers, in water or ice tanks or to other storage materials and uses this stored energy to reduce peak-time electricity consumption for building heating or air conditioning system.

Fig. 4.3 Concentrated solar power with molten salt as heat storage [8]

4.2.6 Pumped Hydroelectric Storage (PHS)

It is a large-scale ESS that uses potential energy of water developed by the gravitational force by pumping water from a lower reservoir to an upper reservoir during low demand time as shown in Fig. 4.4. During high demand time, water is released back into the lower reservoir through turbine to produce electricity. Low energy density of PHS requires either large water body or greater height variation. PHS provides critical backup during peak demand on the national grid.

Power capacity (W) of PHS is a function of the water flow rate and the hydraulic head, while the energy stored (Wh) is a function of the reservoir volume and hydraulic head. Power output of a PHS facility can be calculated by Eq. (4.6) [9] and storage capacity of PHS can be calculated by Eq. (4.7) [10]:

$$P_c = \rho g Q H \eta \qquad (4.6)$$

$$S_c = \frac{\rho g H V \eta}{3.6x10^9} \qquad (4.7)$$

where P_c is the power capacity in Watts (W), ρ is mass density of water (kg/m^3), g is the gravitational constant (m/s^2), Q is the discharge through the turbines (m^3/s), H is the effective head height (m) and η is the generating efficiency, S_c is the storage capacity in megawatt-hour (MWh), V is the volume of water that is drained and filled each day (m^3).

PHS is the cost-effective large storage system currently available although installation requires specific geographical site. An example of PHS is operated by First Hydro Company in UK [11] and the Dinorwig Power Station is capable of moving from 0- to 1,320-MW power injection in 12 s. This station can inject 1,728 MW for 5 h [12]. PHS has comparatively longer life span and can respond quickly to support demand. It is ideal for load levelling applications. It is also suitable for peak load support and frequency regulation.

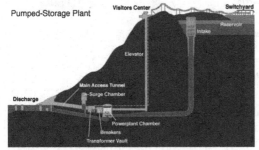
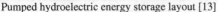

Pumped hydroelectric energy storage layout [13]   PHS facility at Alaska [14]

Fig. 4.4 Pumped hydroelectric storage layout and example

4.2.7 Compressed Air Energy Storage (CAES)

CAES stores energy as compressed air for later use. It consists of a power train motor that drives the compressor, a high-pressure turbine (HTP), a low-pressure turbine (LPT) and a generator as shown in Fig. 4.5. Most commercially implemented CAES systems use diabatic storage system to manage heat exchange. CAES uses off-peak electricity to compress air to store and release compressed air to operate gas turbine. Gas turbine uses compressed air with natural gas therefore efficiency improves using CAES compared to the conventional gas turbine system. Commercial systems use natural caverns as air reservoirs and installed commercial system capacity ranges from 35 to 300 MW.

CAES are considered for applications such as electric grid support for load levelling [7], frequency regulation, load following and voltage control. It is dependent on the specific geographical location for underground reservoir therefore installation cost is high.

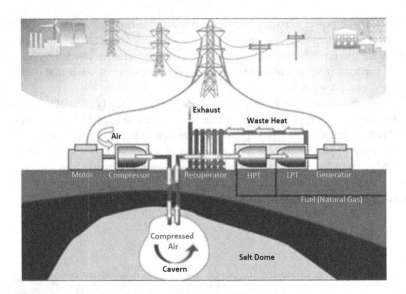

Fig. 4.5 Compressed air energy storage facilities [15]

4.2.8 Hydrogen Energy Storage (HES)

HES differs from the conventional idea of ES, because it uses separate processes for hydrogen production, storage and use. An electrolyzer produces hydrogen and oxygen from water by introducing electric current. A hydrogen fuel cell converts hydrogen and oxygen back into water and release energy. Main drawback of HES is hydrogen is extremely flammable and difficult to store as gas under pressure.

Different strategies of integrating HES with wind and solar energy were proposed in [16].

Fuel cell used stored hydrogen and passed it over the anode (negative) and oxygen over the cathode (positive), causing ions and electrons to form at the anode. The electrons flow through an external circuit that produces electricity while the hydrogen ion passes from the anode to cathode and combines with oxygen to produce water. Figure 4.6 shows the structure of a hydrogen fuel cell.

The characteristics of different ESS described above are summarized in Table 4.1.

In order to support the applications that require combination of high power (for devices with quick response) and high energy (for devices with slow response), hybrid ESS was proposed in different studies [7]. The following section illustrates the applications and benefits of ESS.

Fig. 4.6 Structure of fuel cell

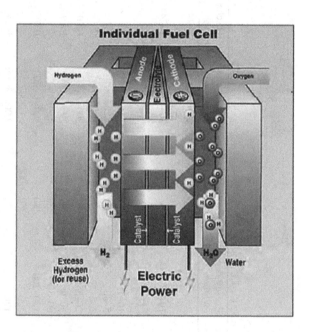

4.3 Applications of Energy Storage System

ES could bring the revolution in electric power system with RE by supporting peak load problem, improving stability and power quality. Storage can be applied with the generation, transmission, in various point of the distribution system, at the customer site or with any particular appliances. ESS in combination of advanced power electronics applies with the intermittent RE sources provides technical benefit, financial benefit and environmental benefit which are explained below.

Table 4.1 Energy storage system [4, 7, 12, 13, 17, 18]

Type	Energy efficiency (%)	Energy density (Wh/kg)	Power density (W/kg)	Life (cycles or years)	Discharge at rated capacity (h)	Response time (s)	Self-discharge
Pumped hydroelectric	70–80	0.3	–	20–60 years	1–24+	10	Negligible
CAES	40–50	10–30	–	20–40 years	1–24+	360	Low
TES	75	–	–	30 years	–	>10 s of minutes	–
SMES	90	10–75	–	>100,000	2.7×10^{-7}–0.0022	0.01	10–15 %
Flywheel (steel)	85–95	5–30	1,000	>20,000	2.7×10^{-7}–0.25	0.1	Very high
Super capacitor	80–95	2–5	800–2,000	10 years	2.7×10^{-7}–1	0.01	5–20 %
Lead-acid	65–80	20–35	25	200–2,000	0.0027–2+	<1/4 cycle	Low
Ni–Cd	60–90	40–60	140–180	500–2,500	0.0027–2+	<1/4 cycle	0.2–0.3 %
Li-MH	50–80	60–80	220	<3,000	–	–	High
Li-ion	70–85	100–200	360	500–10,000	0.017–2+	<1/4 cycle	1–5 %
Li-polymer	70	200	250–1,000	>1,200	–	–	Medium
NaS	70–89	120	120	2,000–3,000	0.0027–2+	<1/4 cycle	–
VRB	80–85	25	80–150	>16,000	0.0027–10	<1/4 cycle	Negligible
EDLC	95	<50	4,000	>50,000	–	–	Very high
Hydrogen	50	100–150	–	–	–	360	Low
Fuel cell	–	–	–	>1,000	0.0027–24+	<1/4 cycle	–

4.3.1 Technical Benefit of ESS

ESS can improve performance of the application suitable for eclectic utility and transport system such as EV. The main advantage of the ESS is to maintain the grid power in constant level [17] by contributing in the following ways.

- *Grid voltage support*:

It means power provided by the ESS to the grid to maintain acceptable grid voltage range especially when RE is integrated.

- *Grid frequency support*:

It means real power provided by the ESS to the grid to reduce any sudden large generation imbalance and to keep the grid frequency within allowable limit. Storage plays significant role in minimizing high-frequency power fluctuations [18].

- *Transient stability*:

ESS reduces power oscillation by injecting or absorbing real power.

- *Load levelling or peak shaving*:

Storing electricity during low demand time and supply it during peak demand time are termed as load levelling as shown in Fig. 4.7. ESS provides freedom in load levelling. Peak shaving moves peak demand into off-peak period.

Fig. 4.7 Basic concept of load levelling using ESS [7]

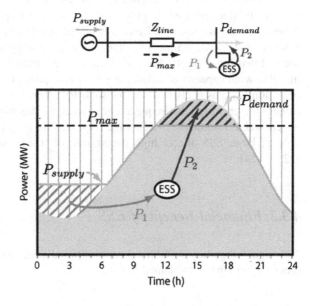

• *Spinning reserve*:

It is defined as the amount of generation capacity that can be used to produce active power over a given period of time.

• *Power quality improvement*:

It is basically the change in magnitude or shape in voltage or current which includes harmonics, power factor, transients, flicker, sag, swell and ESS can mitigate these problems.

• *Reliability*:

It is the percentage or ratio of interruption in electric power delivery to the consumer during total uptime. ESS can reduce the interruption and improve reliability.

• *Ride through support*:

It means electric load stays connected during system disturbance such as voltage sag or momentary blackout. ESS provides support by providing necessary energy to ride through.

• *Unbalanced load compensation*:

ESS can inject or absorb power to/from individual single-phase unbalanced loads. ESS needs to be connected with 4-wire inverter to support in this situation.

• *Increasing penetration of RE sources*:

The intermittent characteristics of solar and wind energy causes fluctuation in voltage and frequency which poses a great barrier in large-scale integration of this fastest-growing RE sources. Moreover, unbalance is demand and supply becomes eminent due to the nature of solar and wind energy generation. Investigation showed that for every 10 % of wind energy penetration into the grid, a 2–4 % of installed wind capacity of balancing power is required from other source for stable operation [7], this is also critical for solar PV integration. ES acts as a buffer that isolates the grid from the frequent and rapid power fluctuations by high penetration of renewable resources [19]. Storage improves the grid penetration of PV energy [20]. A large ESS allows high penetration of wind and solar PV into the grid [12, 21–24].

4.3.2 Financial Benefit of ESS

Although integration of ESS incurs additional cost, however, there are various financial benefits as explained below:

- *Cost reduction*:

Electricity can be purchased during low demand time to store and use it during high demand time, so the overall total consumption cost is reduced. Moreover, stored electricity can be sold during high demand time. Electricity from RE can be used in similar fashion to reduce total consumed energy cost.

- *Avoiding additional cost in generation*:

ESS can help in avoiding installation or renting cost of additional generation to support peak load demand.

- *Avoiding additional cost in transmission/distribution*:

ESS can improve the transmission and distribution performance by operating utilities with its capacity and avoids additional cost of installation to support peak load. Moreover, transmission access/congestion cost can be avoided by the use of ESS.

- *Reduce reliability and power quality–related financial loss*:

ESS helps to improve power quality by supporting loads during outage, sag, and flickering that improves the reliability and that reduces penalty cost of the utility operators.

- *Increases revenue from RE generation*:

ESS helps in time shift in load demand by storing electricity from RE generators and supplying when needed. This ensures maximum utilization of RE.

4.3.3 Environmental Benefit of ESS

- *Reduction in GHG emission*:

ESS helps in best utilization of RE which also reduces the use of conventional energy source and therefore reduces GHG emission [24].

Therefore, ES technology can play a significant role in maintaining power quality and system reliability [2]. The principle application is to respond to sudden changes in load, support load during transmission or distribution interruptions, correct load voltage profiles with rapid reactive power control that allow generators to operate in balance with system load at their normal speed [2]. The technical advantages of different ESS are summarized in Table 4.2.

Proper utilization of ESS depends on various cost involvement with different applications. Cost is one of the key indexes in proper choice of ESS.

Table 4.2 Applications of different energy storage technologies [17, 25]

Energy storage applications	Pumped hydroelectric	CAES	SMES	Lead-acid battery	Flow batteries	Flywheels	Super capacitors	Hydrogen/fuel cell	TES
Load levelling	✓	✓		✓	✓			✓	✓
Load flowing	✓	✓		✓	✓			✓	✓
Peak generation	✓	✓		✓	✓			✓	✓
Fast-response spinning reserve	✓	✓		✓	✓	✓	✓	✓	
Conventional spinning reserve	✓	✓		✓	✓	✓		✓	
Emergency backup	✓			✓	✓	✓		✓	✓
Uninterruptible power supply			✓	✓	✓	✓	✓	✓	
Transient and end-use ride through			✓	✓	✓	✓		✓	✓
Transmission and distribution stabilization		✓		✓	✓	✓		✓	
RE integration	✓	✓		✓	✓			✓	✓
RE backup	✓	✓		✓	✓				✓

4.4 Cost of Energy Storage System

Selection of suitable ESS is determined by the criteria includes lifetime, life cycle, power and energy, self-discharge rate, environmental impact, efficiency, capital cost, storage duration and technical maturity of the storage system. Present cost of storage is considered the major barrier for large-scale utilization. The operational cost (maintenance, loss during operation and ageing) and capital investment cost are the most important factors to select the suitable ESS. Efficiency and lifetime also affect on the overall cost of the ESS. Table 4.3 summarizes the power- or energy-related cost of various ESSs.

Per cycle cost of ESS is 5–80 c/kWh for VRB, 8–20 c/kWh for NaS, 20–100 c/kWh for NiCd and 20–100 c/kWh for lead-acid battery [26].

Table 4.3 Cost of various ESS

Storage technology	Power-related cost ($/kW) [25]	Energy-related cost ($/kWh) [25]	Power capacity cost ($/kW) [4]	Storage cost rank
Pumped hydroelectric	600–2,000	0–20	5–100	Low [13, 26]
CAES	425–480	3–10	2–50	Low [26]
Lead-acid battery	200–580	175–250	50–400	Low [13, 26]]
Ni–Cd battery	600–1,500	500–1,500	400–2,400	High [13, 26]]
NaS battery	259–810	245	300–500	Medium
Li-ion battery	–	900–1,300[13]	600–2,500	High [26]
Vanadium redox	1,250–1,800	175–1,000	150–1,000	Medium
ZnBr	640–1,500	200–400	150–1,000	Medium
High-speed flywheel	350	500–25,000	300–25,000	High [13]
Super capacitors	300	20,000–82,000	300–2,000	High [13, 23]
Fuel cell (hydrogen)	1,100–2,600	2–15	425–725	Low
SMES	300	2,000	1,000–10,000	High [12, 26]

4.5 Classification of Energy Storage

ES will play unique role in future smart grid development by combining different RE sources capability into the grid. Storage can buffer the power spikes and dips and fluctuations [18]. Highest-valued applications of storage identified by EPRI [27] are to maintain commercial and industrial power quality and reliability, to enable stationary and transportable system for grid support.

In large-scale application, electrical ES can be divided into three main functional categories such as follows:

- *Power Quality*: Stored energy applied only for seconds or less to ensure continuity of power quality.
- *Bridging power*: Stored energy applied for seconds to minutes to assure continuity of service when switching from one source of energy to another.

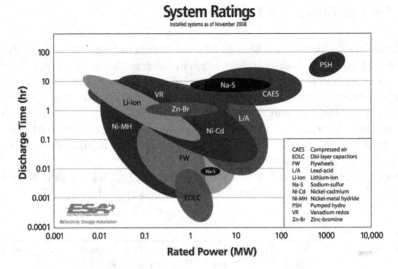

Fig. 4.8 Ratings of different storage systems [28]

- *Energy Management*: Stored energy used to decouple the timing of energy generation and consumption especially in the application of load levelling. Load levelling involves charging of storage in low demand time and uses in peak time which enables consumers to minimize the total energy cost.

ESS has a number of applications in electrical power systems, especially integrating intermittent RE into the grid. Integration of ESS into the grid is briefly explained in Sect. 4.6. ESS can be classified according to capacity, discharge time,

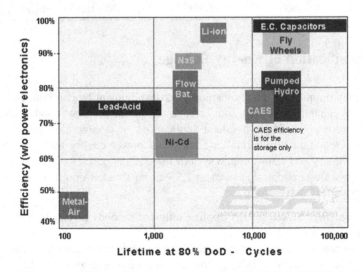

Fig. 4.9 Efficiency and lifetime of each storage technology (at 80 % DoD) [28]

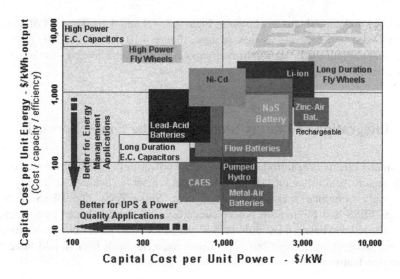

Fig. 4.10 Capital COE storage technology [28]

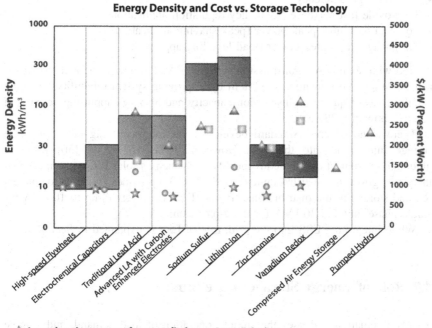

Fig. 4.11 Storage technology and cost comparison [28]

efficiency and capital cost as shown in Figs. 4.8, 4.9 and 4.10. The cost of storage technologies is decreasing as they mature. Figure 4.10 shows the COE based on 2002 value. Figure 4.11 shows the comparison of storage technology to the cost as a function of application.

4.6 Integration of Energy Storage into the Power Network

The main objectives of introducing ESS to the power utility are to improve system load factor, peak shaving, to provide system reserve, achieve reliability and effectively minimize energy production cost. Some ESSs like BESS, EDLC, SMES, FESS and FC require power converter to connect between two different DC voltage level buses, a DC voltage bus and an AC voltage bus or even connect a current source to a voltage bus [7]. Power converter with ESS should have the following features:

- To manage the energy flow in bidirectional way and controlling the charging and discharging process of ESS
- To have high efficiency
- To provide fast response (frequency regulation applications)
- To stand with high peak power (peak shaving application)
- To manage high-rated power (load levelling application)

ESS with flexible AC transmission systems (FACTS) devices adds flexibility to achieve improved transmission system by improving system reliability, dynamic stability, power quality, transmission capacity and also by supporting active and reactive power[2, 29].

There is no concrete standard developed yet for integrating bulk and large storage integration into the grid. However, IEEE 1547-2003 [30] provides guideline to connect distributed resources (DR) such as solar PV, wind and ES to the power grid at the distribution level. AS 4777 [31] provides guideline to connect RE and storage to the distribution network (DN) via inverters up to 10 kVA for single-phase unit and 30 kVA for three-phase unit.

While integrating ESS into the grid, role of ESS is explained in Sect. 4.7.

4.7 Role of Energy Storage: Case Studies

Energy fluctuation collapses the smooth operation of load demand, and storage supports the load by minimizing the fluctuation level. Integration of fluctuating RE into the large flexible grid may not have significant impact; however, if the penetration of fluctuating RE continues to grow, at some point, flexibility of present grid may be fully tapped [32]. Again during best weather condition when RE provides maximum output and during peak time when demand reaches very high, that could potentially breaches the limit of DT capacity and may need to upgrade the costly network capacity. Storage can support the peak load demand and can reduce the load on DT which helps in reducing the overall cost of consumed energy. Storage helps in maximizing the use of RE which eventually helps in reducing greenhouse gas (GHG) emission. Two case studies below show how storage can play role in these regards.

4.7.1 Case Study 1: Storage Role on DT Loading and Minimizing Fluctuations

The case study describes the storage role on DT loading considering residential load. It was also considered that roof-top solar PV was connected to the DN and total load appears on the secondary side of DT.

In order to investigate the storage role, a model was developed in PSS SINCAL as shown in Fig. 4.12. In this model, it was considered that three groups of residential houses connected to the single-phase line of DN through DT. In Australia, residential peak load demand is 1.72 kW [33] or 1.91 kVA considering power factor of 0.9. It was also considered that 5 such houses connected in each phase in each node with same load for each case. Therefore, 15 houses are connected in

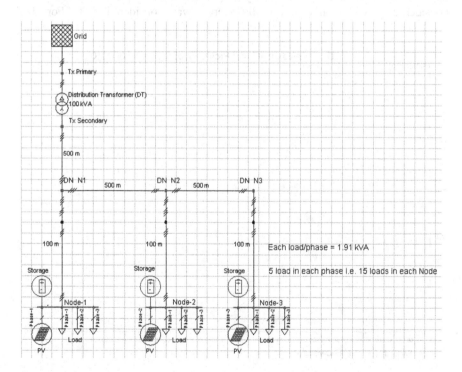

Fig. 4.12 Model scenario

each node and total 45 houses are connected in three nodes from the DT with a total load of 85.95 kVA. It was also considered that 5 such houses installed roof-top solar PV in each node and connected to phase-1 in node-1, phase-2 in node-2 and phase-3 in node-3 as shown in Fig. 4.12. All 3 nodes are considered 500 m apart from each other.

In Queensland, Australia, peak demand generally occurs between 4:00 PM to 8:00 PM when most householders return home and turn on energy-intensive appliances [1]. Daily household load profile is based on the working nature of the residents and the average load pattern as shown in Fig. 4.13.

Urban area load with DT (11 kV/415 V) capacity of 100 kVA was considered in this model. Due to total load and line impedance, DT was 87.22 % loaded in peak demand time (at 20:00 PM) in load-only configuration of the model. Load allocations are shown in Table 4.4.

Daily average residential load and solar radiation in Rockhampton was considered in selecting required PV capacity. Ergon Energy, local distribution network service provider (DNSP) in Rockhampton, Australia, allows 4-kW-capacity PV for each urban area house [34]. The output from PV is not available for 24-h period, and residential load demand is lowest when PV generates highest energy and excess energy supplies to the grid eventually increases the voltage at the connected node. In Australia, yearly average sunlight hours vary from 5 to 10 h/day and maximum area is over 8 h/day [35]. In this model, inverter efficiency was considered as 97 % and loss until inverter was considered 5 %. For this

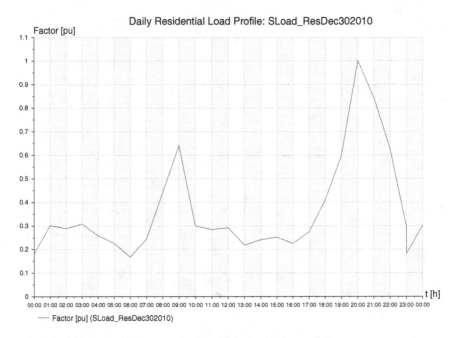

Fig. 4.13 Daily residential summer load profile in Rockhampton [24]

Table 4.4 Load allocation in three nodes

Node	Phase 1		Phase 2		Phase 3	
	kVA	cos φ	kVA	cos φ	kVA	cos φ
Node 1, 2, 3	9.55	0.9	9.55	0.9	9.55	0.9

Table 4.5 Installed solar PV in different nodes

Node	Phase 1		Phase 2		Phase 3	
	kW	No of house	kW	No of house	kW	No of house
Node 1	5	5	–	–	–	–
Node 2	–	–	5	5	–	–
Node 3	–	–	–	–	5	5

Table 4.6 Installed storage in different nodes

Node	Phase 1		Phase 2		Phase 3	
	kW	No of house	kW	No of house	kW	No of house
Node 1	1.72	5	–	–	–	–
Node 2	–	–	1.72	5	–	–
Node 3	–	–	–	–	1.72	5

investigation, it was considered that five houses in node-1 installed 5-kW PV/ house in phase-1, similarly 5 houses in node-2 and 5 houses in node-3. It was also considered that storage was installed with the same peak capacity of load which is 1.72 kW in each house where solar PV was installed. Tables 4.5 and 4.6 show the installed PV and storage in different nodes.

Daily utilization of DT in supporting the allocated load is shown in the Fig. 4.14, and the peak utilization was found 87.22 % of DT capacity at the time of 20:00 PM.

It was explained earlier that solar PV was added to phase-1 in node-1 and summertime solar radiation in Rockhampton was considered for the analysis. Daily solar radiation data of Rockhampton was collected from [35] and daily profile is as shown in Fig. 4.15. However, due to cloud movement, energy level received by the PV array fluctuates and fluctuation varies with the variation of cloud movement. Figure 4.16 shows the dip fluctuation in solar radiation for a long period.

Simulation was conducted for load flow (LF) and load curve (LC) analysis. LF is an effective tool for calculating the operational behaviour of electrical transmission and distribution network. LF calculates current and voltage distribution from generation to the consumption on rated power or voltage at the node elements. LC is a LF calculation with load values varied over time.

It was found from LC simulation that PV increases the DT loading during peak generation time (especially at 13:00 PM) when load demand was low and when storage was not integrated with the system. However, maximum loading was at 20:00 PM when residential load demand is highest, and when PV was not able to

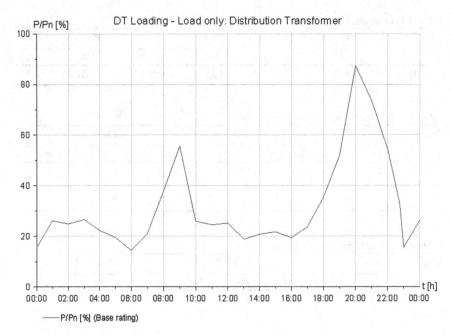

Fig. 4.14 DT loading due to load only

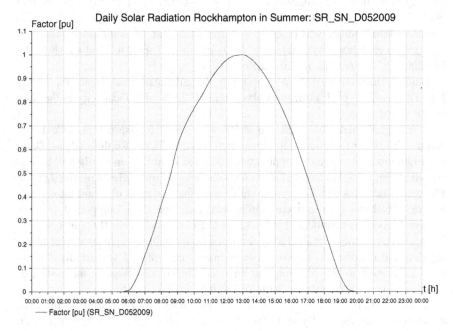

Fig. 4.15 Daily solar radiation profile of Rockhampton, Australia

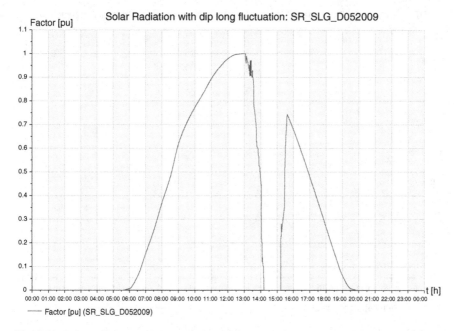

Fig. 4.16 Solar radiations with long dip fluctuation

support the load, therefore maximum load on DT remains 87.22 % as shown in Fig. 4.17. After integrating PV in all 3 nodes, three phases in 3 nodes become balanced in loading and support the load in the morning which lowers the DT loading in the morning at 08:00 AM and afternoon at 18:00 PM as shown in Fig. 4.17. However, PV increases DT loading during midday (52.06 % at 13:00 PM) from 10:00 AM to 16:00 PM and also peak load in the evening remains same as 87.22 %.

When storage was integrated only in node-1 and connected to phase-1, storage supported phase-1 load although which was not enough compared to the total load on phase-1; therefore, it was not reflected on overall DT loading. By adding storage in node-2 and connecting to phase-2, it was found that maximum/peak loading on DT now reduced to 80.51 %. Moreover, midday loading also reduced to ~42 % at 13:00 PM. Gradually, storage was added in phase-3 in node-3 and found that loading on DT reduced not only during midday but also in peak demand time in the evening. Figure 4.18 shows the charging and discharging period of storage connected in phase-3 in node-3.

After integrating storage in all three phases, it was found that peak-time loading reduced to 66.76 % as depicted in Fig. 4.19 which is a great improvement in RE utilization and also reduced the risk of upgrading the DT capacity. Moreover, storage also reduced the load on DT during day time when residential load demand was low particularly during 10:00 AM to 16:00 PM, and during this time, maximum loading reduced to ~30 % of DT capacity at 13:00 PM.

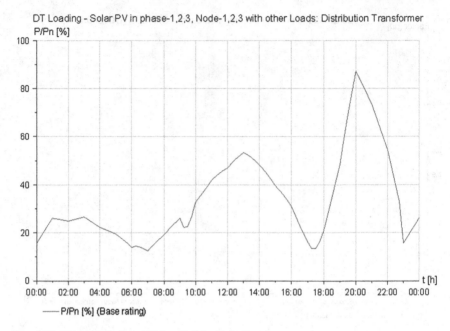

Fig. 4.17 DT loading when PV installed in all 3 phases

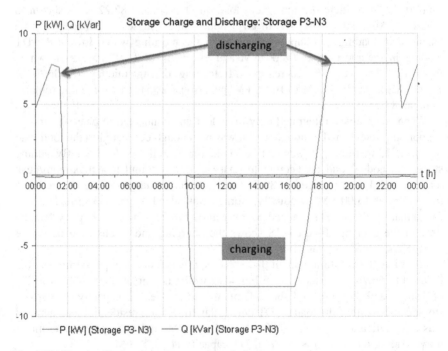

Fig. 4.18 Charging/discharging of storage in node-3

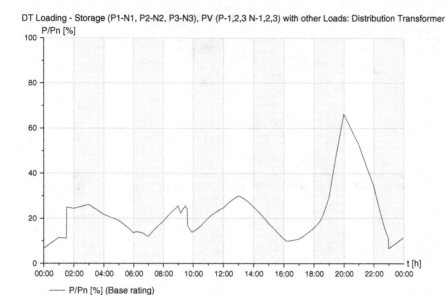

DT Loading - Storage (P1-N1, P2-N2, P3-N3), PV (P-1,2,3 N-1,2,3) with other Loads: Distribution Transformer

Fig. 4.19 DT loading after adding storage in all 3 phases

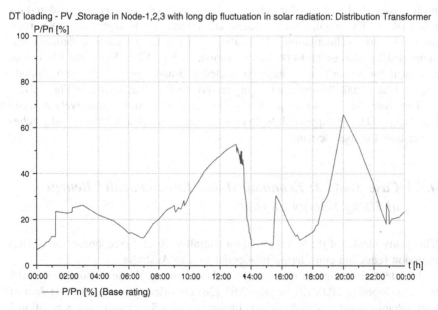

DT loading - PV _Storage in Node-1,2,3 with long dip fluctuation in solar radiation: Distribution Transformer

Fig. 4.20 DT loading when PV and storage connected and solar radiation with long dip fluctuation

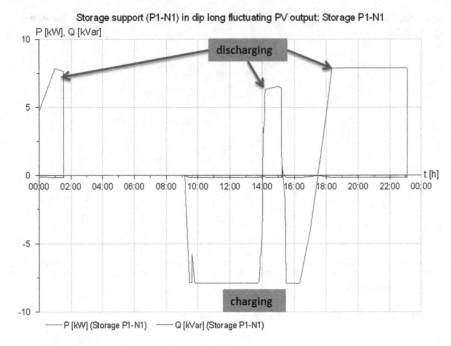

Fig. 4.21 Storage supports load when solar PV output has long dip fluctuation

Cloud movement or various natural conditions could change the solar radiation profile from ideal type to solar radiation with long dip fluctuation as considered for this analysis. PV output also fluctuates due to long dip fluctuations in solar radiation. Due to the fluctuation, DT loading also impacted, and load demand was supported by storage at 14:00 PM as shown in Fig. 4.20. While the PV output interrupts for a long time, storage connected in 3 nodes in 3 phases also supports the load (at 14:00 PM) by discharging stored electricity as shown in Fig. 4.21.

Therefore, this investigation clearly illustrated that storage effectively reduced the load on DT and supported the load when RE generation fluctuates and reaches lower than the load demand.

4.7.2 Case Study 2: Economical and Environmental Benefit of Using Energy Storage

This study identified the storage role on overall COE and greenhouse gas (GHG) emission reduction considering residential load in Australia.

In order to investigate the economic and environmental role of storage, a model was developed in HOMER version 2.68 [36] considering solar and wind data of two potential locations in Australia. Integration of ES certainly incurs additional cost to the system; however, study showed that storage increases RE utilization,

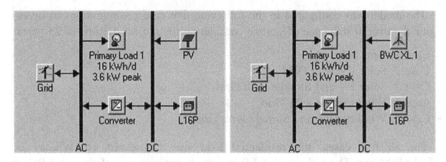

Fig. 4.22 Model configurations (Grid-Solar-Storage and Grid-Wind-Storage)

reduces COE and reduces GHG emission [37]. Residential load was considered for this analysis, and daily load profile is shown in Fig. 4.13. Model configuration is shown in Fig. 4.22. The costing of all required components was considered according to the available market price of each unit. Grid electricity price for Tariff-11 in Queensland is $0.3145/kWh (including GST and service) [38]. According to the load demand, grid electricity cost was considered for off-peak, peak and super-peak period. Element cost and supporting considerations of the required components are shown in Table 4.7.

Table 4.7 Technical data and study assumptions

Description	Cost/information
PV array [39]	
Capital cost	$3,100.00/kW
Replacement cost	$3,000.00/kW
Operation and maintenance cost	$50.00/year
Lifetime	25 years
Wind turbine [40]	
Capital cost	$4,000.00/kW
Replacement cost	$3,000.00/kW
Operation and maintenance cost	$120.00/year
Lifetime	25 years
Grid electricity [41]	
Off-peak rate (09:00 AM–06:00 PM, 10:00 PM–07:00 AM)	$0.30/kWh
Peak rate (07:00 AM–09:00 AM, 08:00 PM–10:00 PM)	$0.35/kWh
Super-peak rate (06:00 PM–08:00 PM)	$0.45/kWh
Inverter [42]	
Capital cost	$400.00/kW
Replacement cost	$325.00/kW
Operation and maintenance cost	$25.00/year
Lifetime	15 years
Storage (battery) [43]	
Capital cost	$170.00/6 V 360 Ah
Replacement cost	$130.00/6 V 360 Ah
System voltage	24 V

The model was configured in the following five case configurations to investigate the overall influence of storage considering the project lifetime of 25 years.

- Case 1: Grid only
- Case 2: Solar PV with Grid
- Case 3: Solar PV and Storage with Grid
- Case 4: Wind turbine with Grid
- Case 5: Wind turbine and Storage with Grid

For all five cases, the residential load was considered 16 kWh/day or 5,840 kWh/year. The influence of ES in overall performance of the solar PV or wind turbine system was analysed by evaluating GHG emission and the COE of the present system (Case-1) and after integrating storage with the solar PV or wind turbine integrated system. Australia is one of the best places in the world for solar and wind energy, and Alice Springs is one potential location in Queensland for solar energy where as Macquarie island in Tasmania has great potential for wind energy [24]. Optimization result showed that for a residential load of 16 kWh/d in Alice Springs solar PV generates electricity that reduces GHG emission by 18.69 % (Case-2). However, after adding battery as storage with the system, GHG emission reduces 79.05 % (Case-3). This significant reduction in GHG emission achieved as storage increased the use of RE, and without storage, this energy was wasted and load demand was supported by conventional sources. Similarly Wind turbine in Macquarie Island reduces 59.19 % GHG emission (Case-4) for the same load; however, after adding storage, much more electricity was consumed from wind source and enough electricity was sold back to the grid that GHG emission reduced up to 167.78 % in Case-5. Figure 4.23 shows the total GHG emission to support the residential load in five case configurations.

Optimization results showed that to support 5,840 kWh/year of residential load in Alice Springs, 1-kW PV with 1-kW inverter was used (Case-2) without battery and only 26 % of this load was supported by PV while remaining load was supported by the grid; therefore, overall COE becomes $0.376/kWh. However, in storage-integrated system (case-3) for the same load at the same place, 3-kW PV was used with 2-kW inverter and 16 batteries. This optimized model supports

Fig. 4.23 Greenhouse gas (GHG) emission in different configurations

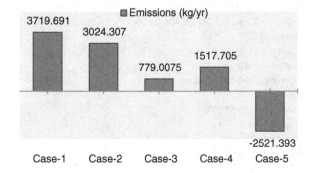

Fig. 4.24 COE in different configurations

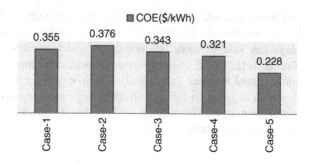

75.51 % of load and storage supports the load for extended period that reduces use of grid electricity and overall COE becomes $0.343/kWh as shown in Fig. 4.24. At DC system voltage of 24 V, this configuration needs 1440 Ah of battery support. In Macquarie Island, 1-kW wind turbine with 1-kW inverter supports 30.21 % of load (Case-4) without storage and generates much more electricity than required; as a result, 31.39 % generated electricity was wasted and overall electricity cost becomes $0.321/kWh. When battery was added to support, optimized model used 2-kW wind turbine, 5-kW inverter and 12 batteries. This configuration supports 96.42 % of load demand and increased electricity export to grid and overall COE becomes $0.228/kWh. To support 96.42 % of load, this configuration used 1,080-Ah battery at 24-V system voltage.

Therefore, storage has strong influence in greater utilization of RE and that reduces GHG emission and COE.

4.8 Future of Energy Storage and Conclusions

Although there are currently few installations of large-scale ES exists, the potential of ESS and advantages with various applications have been noted in different research findings. Future use of ES is concentrating mostly with stationary energy and transport sector. Research forecasted utility ES market will grow from $329 million in 2008 to approximately $4.1 billion in 10 years [44]. Cost of ES is considered the main barrier for wider use [32]. Research is going on for efficient, cost-effective storage for grid integration [45]. The importance of ES has been evaluated by forming working group to develop standard IEEE P2030.2 for the interoperability of ESS integration with the electric power infrastructure [46]. Advantages of ES already proved its potential in utility and transport sectors.

This chapter described various storage systems and their characteristics that show their potential to be used in various power quality application, load levelling, load shifting, transmission/distribution system stabilization and RE integration application. This chapter also showed the cost comparison among different storage systems. Case studies showed that storage significantly improves the capability of DT by reducing loading on it which eventually reduces the risk of upgrading DT and transmission/distribution capacity. Storage reduces fluctuations generated by RE and support the load during fluctuation. Other case study showed that storage plays key role in more RE utilization which eventually reduces overall COE. Storage helps in RE utilization that reduces consumption of grid electricity from conventional sources (e.g., coal) which ensures reduction in GHG emission. Research is going on for cost-effective large storage, and investment is increasing in storage market. Therefore, it is likely that storage will play the key role in future power network applications.

References

1. Queensland Energy Management Plan, Department of Employment, Economic Development and Innovation, Queensland Government (2011) Available [Online] at: http://rti.cabinet.qld.gov.au/documents/2011/may/qld%20energy%20management%20plan/Attachments/Qld%20Energy%20Mgt%20Plan.pdf. Accessed 28 Aug 2012. Report, May 2011
2. Ribeiro PF et al (2001) Energy storage systems for advanced power applications. Proceedings of the IEEE, December 2001. 89(12):12
3. Miller NW et al (1996) Design and commissioning of a 5 MVA, 2.5 MVA battery energy storage. Proceedings of IEEE power engineering society transmission and distribution conference, 1996
4. Beaudin M et al (2010) Energy storage for mitigating the variability of renewable electric sources: an updated review. Energy for Sustainable Development, Elsevier, 2010 14(4):302–314
5. Oudalov A et al (2006) Value analysis of battery energy storage applications in power systems. Proceedings of PSCE-2006, IEEE, 2006
6. Abi-Samra N. et al (1995) The distribution system dynamic voltage restorer and its applications at industrial facilities with sensitive loads. Proceedings of 8th international power quality solutions, Long Beach, CA, 1995
7. Vazquez S et al (2010) Energy storage systems for transport and grid applications. IEEE Transactions of Industrial Electronics, 2010. 57(12):3881–3895
8. RGEP (2012) Renewable green energy power, solar energy facts—solar energy storage enables CSP and PV on the grid. Available [Online] at: http://www.renewable greenenergypower.com/solar-energy-facts-solar-energy-storage-enables-csp-and-pv-on-the-grid/ Accessed 05 Nov 2012
9. Wong IH (1996) An underground pumped storage scheme in the Bukit Timah granite of Singapore. Tunnelling and Underground Space Technology, Elsevier, 1996. 11(4):485–489
10. Figueiredo FC, Flynn PC (2006) Using diurnal power price to configure pumped storage. IEEE Transactions on Energy Conversion, 2006. 21(3):804–809
11. FHC (2012) Pumped hydro storage, first hydro company. Available [Online] at: http://www.fhc.co.uk/index.asp. Accessed 04 Nov 2012
12. Diaz-Gonzalez F, Sumper A, Gomis-Bellmunt O (2012) A review of energy storage technologies for wind power applications. Renew Sustain Energy Rev 16(4):2154–2171
13. Hadjipaschalis I, Poullikkas A, Efthimiou V (2009) Overview of current and future energy storage technologies for electric power applications. Renewable and Sustainable Energy Reviews, Elsevier 2012. 13(6–7):1513–1522
14. ACEP (2012) Energy storage, Alaska energy wiki. Available [Online] at: http://energy-alaska.wikidot.com/energy-storage. Accessed 04 Nov 2012
15. Compressed air energy storage (CAES) in salt cavern. Available [Online] at: http://web.evs.anl.gov/saltcaverns/uses/compair/index.htm Accessed 10 Nov 2012

16. Korpass M, Greiner C (2008) Opportunities for hydrogen production in connection with wind power in weak grids. Renewable Energy, 2008. 33(6:1):199–208
17. Oudalov A, Buehler T, Chartouni D (2008) Utility scale applications of energy storage. In: Proceedings of Energy 2030, IEEE. Atlanta, Georgia, USA, 2008
18. Hittinger E, Whitacre JF, Apt J (2010) Compensating for wind variability using co-located natural gas generation and energy storage. Energy System, Springer 2010. pp 417–439
19. Makarov Y. et al (2010) Optimal size of energy storage to accommodate high penetration of renewable resources in WECC system. Proceedings of IEEE, 2010
20. Solomon AA, Faiman D, Meron G (2010) Properties and uses of storage for enhancing the grid penetration of very large photovoltaic systems. Energy Policy, Elsevier, 2010. 38:5208–5222
21. Mohod SW, Aware MV (2008) Energy storage to stabilize the weak wind generating grid. Proceedings of power system technology and IEEE power India conference, POWERCON 2008, pp 1–5
22. Tanabe T et al (2008) Generation scheduling for wind power generation by storage battery system and meteorological forecast. In: Proceedings of power and energy society general meeting—conversion and delivery of electrical energy in the 21st century, IEEE 2008, pp 1–7
23. Ibrahim H, Ilinca A, Perron J (2008) Energy storage systems—characteristics and comparisons. Renewable and Sustainable Energy Reviews, Elsevier, 2008. 12(5):1221–1250
24. Arif MT, Oo AMT, Ali ABMS (2012) Investigation of energy storage required for various location in Australia. Proceedings of central regional engineering conference 2012, Engineers Australia, Queensland, Australia, 2012
25. Connolly D (2010) The integration of fluctuating renewable energy using energy storage. PhD thesis, Department of Physics and Energy, University of Limerick. Available [Online] at: http://www.dconnolly.net/files/David%20Connolly,%20UL,%20Energy%20Storage%20 Techniques,%20V3.pdf. Accessed 03 Nov 2012
26. Rahman F, Rehman S, Abdul-Majeed MA (2012) Overview of energy storage systems for storing electricity from renewable energy sources in Saudi Arabia. Renewable and Sustainable Energy Reviews, Elsevier, 2012. 16(1):274–283
27. Boutacoff D, Rastler D, Kamath H (2010) Energy storage, enabling grid-ready solutions, EPRI. Available [Online] at: http://eepublishers.co.za/article/epri-144-01-energy-storage-enabling-grid-ready-solutions.html. Accessed 28 Aug 2012. Article, 2010
28. ESA (2012) Technology comparisons, energy storage association (ESA). Available [Online] at: http://www.electricitystorage.org/images/uploads/static_content/technology/technology_ resources/ratings_large.gif. Accessed 04 Nov 2012
29. ABB, Connecting renewable energy to the grid. Available [Online] at: http://www.abb.com/ cawp/seitp326/377dbeff7a3a6aedc12577c20033dbf5.aspx. Accessed 01 Nov 2012
30. IEEE (2003) IEEE standard 1547-2003. IEEE standard for interconnecting distributed resources with electric power systems. Standard, 2003, pp 1–16
31. Standard Australia (2005) AS 4777.1-2005 Grid connection of energy systems via inverters, part 1: installation requirements. Available [online] at: http://www.saiglobal.com/. Accessed 18 Jun 2012. Standard, 2005
32. Goggin M (2010) Wind power and energy storage, American wind energy association. Available [online] at: www.awea.org. Accessed 28 Aug 2012. Fact Sheet, July 2010
33. Origin Energy (2012) household peak demand. References available [Online] at: http:// www.originenergy.com.au/3404/Mt-Stuart-Power-Station. Accessed 10 Oct 2012
34. Thomas A (2011) Photovoltaic planning criteria, Network planning and development distribution planning and capability, Ergon Energy, Australia. Technical note, 2011
35. Meteorology, B.o., weather data, Bureau of meteorology, Australian Government. Available [Online] at: http://reg.bom.gov.au/. Data collected on 03 Jun 2011
36. HOMER, Analysis of micro power system options. Available [Online] at: http:// homerenergy.com/software.html Accessed 07 Nov 2010

37. Arif MT, Oo AMT, Ali ABMS (2013) Estimation of energy storage and its feasibility analysis, energy storage—technologies and applications. Zobaa A (ed) , InTech. ISBN: 978-953-51-0951-8, doi: 10.5772/52218. Available from: http://www.intechopen.com/books/energy-storage-technologies-and-applications/estimation-of-energy-storage-and-its-feasibility-analysis. Accessed Jan 2013
38. Ergon Energy, Electricity tariffs and prices. Available [Online] at: http://www.ergon.com.au/your-business/accounts-and-billing/tariffs-and-prices-2012-13 Accessed 12 Oct 2012
39. PV-Price (2012) Goodhew electrical and solar. Available [Online] at: http://www.goodhewsolar.com.au/customPages/goodhew-electrical-%26-solar-offers-homeowners-the-most-affordable-quality-solar-systems-on-the-market.?subSiteId=1. Accessed 15 Mar 2012
40. Winturbine_Price (2012) Ecodirect, Clean energy solution, wind turbine price. Available [Online] at: http://www.ecodirect.com/Bergey-Windpower-BWC-10kW-p/bergey-windpower-bwc-10kw-ex.htm. Accessed 29 Feb 2012. Online Information
41. RedEnergy (2011) Pricing definition for electricity customers, NSW. Available [Online] at: http://www.redenergy.com.au/docs/NSW-Pricing-DEFINITIONS-0311.pdf
42. Inverter-Cost, Sunny Boy 1700 price. Available [Online] at: http://www.solarmatrix.com.au/special-offers/sunny-boy-1700?ver=gg&gclid=CMO0gIOzna4CFYVMpgod7T1OHg Accessed 16 Mar 2012
43. ALTE-Store (2012) Battery price, Trojan T-105 6 V, 225 Ah (20HR) flooded lead acid battery. Available [Online] at: http://www.altestore.com/store/Deep-Cycle-Batteries/Batteries-Flooded-Lead-Acid/Trojan-T-105-6V-225AH-20HR-Flooded-Lead-Acid-Battery/p1771/. Accessed 16 Mar 2012
44. Link D, Wheelock C (2009) Energy storage technology markets. Research report, Pike research, 2009
45. Wessells C (2011) Stanford university news, nanoparticle electrode for batteries could make large-scale power storage on the energy grid feasible. Available [Online] at: http://news.stanford.edu/news/2011/november/longlife-power-storage-112311.html. Accessed 24 Nov 2011
46. IEEE (2012) IEEE P2030.2 Draft guide for the interoperability of energy storage systems integrated with the electric power infrastructure. Available [Online] at: http://grouper.ieee.org/groups/scc21/2030.2/2030.2_index.html. Accessed 05 Nov 2012

Chapter 5
Smart Meter

M. Rahman and Amanullah M. T. Oo

Abstract Utility meters are being changed from simple measurement devices to multi-dimensional technical devices and also enhanced by the addition of new informational and communication capacities like smarter metering systems [1]. Smart meters enable automatic, bi-directional communication between the consumers and the utility. Compared to traditional energy, meters display only the amount of energy consumed, but smart meters can directly send usage data back to the utility. The information of electricity consumption could be collected in real time with accuracy from smart meter [2]. Modern distribution companies are required to adopt smart meters in their network to improve the efficiency of the networks and to be in par with the smart grid environment. This chapter has conducted a rigorous review that outlines the existing distribution network, deployment of smart meter in distribution network, and possible difficulties to deploy smart meter network in distribution system. The purpose of this chapter is to provide the necessary background to understand the concepts related to smart meter, smart meter network, and relevant research carried out in this area. A concise review of importance of implementing smart meter in distribution network, bandwidth requirement for smart meter network, bandwidth and latency barrier in smart meter network, and communication coverage of smart meter network is presented.

M. Rahman (✉)
Queensland University of Technology, Brisbane, Australia
e-mail: m2.rahman@qut.edu.au

A. M. T. Oo
Deakin University, Geelong, Australia

A. B. M. S. Ali (ed.), *Smart Grids*, Green Energy and Technology,
DOI: 10.1007/978-1-4471-5210-1_5, © Springer-Verlag London 2013

5.1 Details Description of Smart Meter

Smart meters are placed in consumer's premises that exchange information between consumers and control center. A smart meter has the following capabilities [3]:

- Real-time data of electricity use and electricity generated locally;
- Offering the possibility to read the meter readings both locally and remotely;
- Offering the interconnection between premise-based networks and devices such as DG.

Usually, a smart meter is considered for registry of electricity use. The intelligence of the meter is incorporated in the smart meter. It has several basic functions, namely measuring the electricity consumed or generated, remotely switching the customer and remotely controlling the maximum electricity used as shown in Fig. 5.1, [3]. Furthermore, the design processes of smart meter create many opportunities across smarter metering technologies [4].

Smart meters refer to systems in which information flows in both directions. These systems open up a wide range of opportunities for utilities through up-to-date information and innovative products. There is a growing interest to know how the new information and communication technologies (ICT) can change the conventional meter into a smart meter for energy saving and energy security purposes [1]. The functionalities of new metering technologies are brought together in distinct packages in this smart metering system.

Direct energy consumption monitoring and improving energy efficiency are the main purpose of smart metering on distribution networks. Smart meters encourage consumers to change behavior by turning down the electrical appliances during peak demand period. These effects are fostered either by exposing consumers to their general consumption patterns or by financial incentives. Energy efficiency can be improved as consumer regulates power consumption based on their needs. The data generated and transmitted by smart meters open up a range of operational

Fig. 5.1 Schematic diagram of a typical smart meter configuration [3]

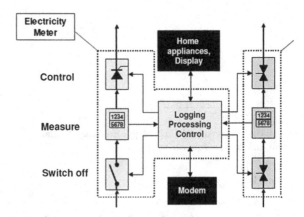

improvements including monitoring and controlling energy usages. This detailed network information improves the network operations and facilitates more renewable energy generation and integration. The broad range of energy saving opportunities allows smart meter an attractive option for distribution network. Benefits of smart meter in distribution network, investigation of the requirements of smart meter in distribution, technological configurations of smart meter, smart meter monitoring program, and impact of smart meter in distribution system are discussed in the following few subsections.

5.1.1 Benefits of Smart Meter

Smart distribution network is required to maintain and enhance energy security. Globally, the indigenous energy resources to secure energy supplies have driven the development and deployment of new and renewable technologies including advancing ways in which energy is delivered by smart meter. The implementation of a smart meter in distribution system can facilitate a step transformation in the way energy is produced and consumed. Smart meter with high-tech communication capabilities monitor energy usage and allow consumers to make informed choices about how much energy to be used and when to consume it [5].

5.1.2 Requirement of Smart Meter in Smart Distribution Network

Smart meter provides consumers with the ability to use electricity more effectively and provides utilities with the ability to operate them more efficiently. Adoption of smart meter in smart distribution network will enhance the electric delivery system including generation, transmission, distribution, and consumption. It allows the possibilities of DG bringing generation closer to the consumers [5]. This shorter distance from generation to consumption empowers consumers to be active participants in their energy choices.

Furthermore, the significance of smart metering also rises from eliminating the need for more energy generation which reduces the requirement of electricity grid expansion to overcome the electricity congestion problem. This is possible by means of distributed generation and demand response [6]. Energy consumption levels are substantially reduced, and peak load generation is lessened as loads are being shifted from peak to off-peak [7]. Smart meters act as the intelligent nodes on smart distribution network. The increasing importance on energy from renewable energy sources puts a further strain on the current distribution network, which is currently "one way" only—it cannot accept from micro-generated electricity. From the operational point of view, the use of smart metering system allows for improved management and control over the electricity distribution system [8]. The smart distribution system should be more reliable, reducing significantly the staggering cost of power outages for consumers and businesses. Today, the tools for improving real-time monitoring and controlling the distribution system with advanced information technology are available. The deployment of smart meter in distribution system with advanced information technology can manage these increasingly complicated networks. An integrated, information technology (IT)-enabled, electric distribution system is required for improving reliability of distribution system, but it is really critical for bringing higher percentages of renewable electricity energy into our energy mix due to the variable nature of many of these resources. By introducing a large number of IEDs (like smart monitors and control points) in the distribution network, the smart distribution network enables more precise management, which can increase effective capacity [8].

5.1.3 Technical Configurations of Smart Meter

Rapid changes in ICT have dramatically changed the potential of utility meters and the new system to measure the energy consumption was known as "smart meter" as a reflection of their increased functionalities and communication capacities compared to the "simpler" predecessors which mainly measure the energy consumption by manual reading [1]. Electric meter manufacturers and information communications providers have been competing to provide utilities with their new smart metering systems.

Current distribution networks are old and were not designed to cope with future electricity demand. Smart meter is also a key element for providing advanced home appliances and end-use technologies on line to capture new efficiencies [8]. From the operational point of view, the use of smart metering system allows for improved management and control over the electricity distribution system. Moreover, the future demand distribution network made available by using smart metering facilitates for further grid planning.

Meter manufacturers and communications providers have been competing to provide utilities with their new smart metering systems. Siemens Limited is one of the smart meter provider organizations, and their devices are high-performance power monitoring devices [3]. The devices are used to detect the power values for electrical feeders or individual consumers. Furthermore, the devices provide important measured values for assessing the system state and the power quality. The devices provide accurate knowledge of your systems characteristics and offers more than 50 basic values regarding the power quality as shown in the Table 5.1.

Table 5.1 Smart meter parameters settings [4]

Function	Parameters/infrastructure performance
Basic measuring values	Voltage, current, active power, reactive power, power factor, frequency
Extended measuring values	Phase displacement angle, Phase angle, harmonics voltage, harmonics current, distortion current, maximum values/date and time, asymmetry voltage
Power detection/ counters	Apparent energy, active energy, reactive energy, power demand of last measuring period, measuring period, operating hours counter
Interfaces	Ethernet, simultaneous connection, protocol, gateway
Input/outputs	Digital input, digital output, operating voltage
Clock/calendar	Real-time clock, calendar function, summer time/winter time switchover
Display/operation	Display, indication, operation, language

5.1.4 Smart Meter Monitoring Program

Smart meter is a two-way information communication system. The basic elements of a smart meter monitoring program are shown in Fig. 5.2. It includes smart meters, a means of communication, and a power quality data warehouse. The means of communication could include fixed twisted pair cable, telephone lines, mobile phones, power line carrier, radio, fiber optic, or a combination of these [9].

The chosen meters must be capable of recording the parameters. The data warehouse must be an effective and secure means of storing data from all meters in the monitoring period. The quantity of data stored will be dependent upon the transmitted data [9]. Present smart meters that capture power quality data have concentrated on transmitting all data such as 10 min data for voltage, current, power factor as well as others data. The data must first be sent from the smart meter to the power quality data warehouse. A more sensible and realistic option is for each smart meter to report by exception in real time or using power quality indices for each site. However, adoption of smart meter that faces several challenges in smart distribution network can be described in the following sub-section.

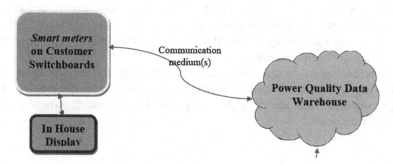

Fig. 5.2 Smart meter monitoring program [9]

5.1.5 Impact of Smart Meter in Distribution System

Despite a number of benefits from the widespread application of smart metering technology, the adoption of this new technology faces technical challenges. The smart meter network faces new challenges and stresses which may put at risk its ability to reliably deliver power. The challenges include the following:

Communication network challenges: Smart meters offer better interface technologies. As smart meters enable bi-directional communication between the customer and the utility, these devices provide access to more information than traditional meter. To facilitate a robust communication between smart meter devices and centralized control system, a high bandwidth communication channel is required to allow digital data move throughout the network efficiently and effectively.

Standard communication protocol: Lack of standardization of smart metering technology means that large number of smart meters of different types has been deployed for collecting and dispatching data and instructions, transforming the data and storing data under different communication protocols [10]. If same communication protocol is used, the problem may be minimized. Otherwise, international standards covering automatic meter data exchange can also overcome this technical barrier [11]. The strategies outlined above provide a bridge to the implementation of future smart technologies where real-time data will be the best practice network. They will provide a basis for implementing smart meter network in distribution system which provides a range of security, quality, and reliability.

5.2 Smart Meter in Distribution Network

Smart meter in distribution network covers modernization of distribution system. This system is directed at several disparate set of goals including facilitating competition between providers, enabling a large-scale use of variable energy sources, establishing the automation, and monitoring capabilities in distribution system [10]. The smart meter network is used to describe the integration of the elements connected to the electrical grid with an information infrastructure to offer numerous benefits for both the providers and consumers of electricity. It is an intelligent future electricity system that demands two-way information systems through an intelligent communication system.

The technologies, devices, and systems that make up a smart distribution network will vary across electricity distribution systems, like as the existing electricity networks vary according to the geographic, climatic, ownership, and business parameters.

5.2.1 Components of Smart Meter Network in Distribution System

The smart meter network components include [11]:

- integrated communications infrastructure that enables near real-time, two-way exchanges of information and power;
- smarter measurement devices (including advanced metering infrastructure) that record and communicate more detailed information about energy usage;
- sensors and monitoring systems throughout the network that keep a check on the flow of energy in the system and the performance of the network's assets;
- automatic controls that detect and repair network problems and provide self-healing solutions;
- advanced switches and cables that improve network performance; and
- IT systems with integrated applications and data analysis.

Figure 5.3 describes new opportunities that are enabled by smart distribution networks, rather than being a component part of smart networks. There are a range of other technologies, devices, and applications that are often associated with smart meter in distribution networks, such as customers' energy management systems, renewable energy supplies, and energy storage technologies [11].

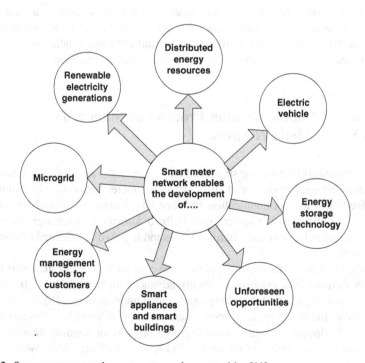

Fig. 5.3 Smart meter network components and opportunities [11]

5.2.2 Characteristics of Smart Meter in Distribution Network

The following provides additional attributes on the smart distribution network characteristics [11–13]:

- active participation of consumers occurs through the provision of two-way communications and information that gives the consumer the ability to consume and provide energy;
- achieve self-healing (i.e., automatic fault response), the integration of devices and sensors with a secure communications network will automatically recover unaffected sections of the network and isolate those elements in need of repair. Resistance to security attack is enhanced as end-to-end cyber-security is enforced across the network with smart network security protocols;
- enable generation and storage options at the macro and micro level by way of participatory networks established at all levels of the network, allowing individual and industrial customers;
- optimize assets and efficient operation and reduce operating and maintenance costs, harnessing the information provided by sensing and monitoring devices and automatic switching capability;
- enable distribution network is flexible to natural disasters with rapid restoration capabilities.

Therefore, it clearly indicates that smart distribution network is a vital need today for a sustainable, secured power system for the future. However, development of smart meter network introduces potential difficulties includes communication network challenges which are discussed in the next section.

5.3 Roles of Communication Protocol or Standard in Smart Meter Network

A communications protocol or standard is a system that has rules for exchanging information between nodes or stations. In smart meter network, this communication facilitates the communication between smart meter to central server and vice versa. Communication system uses defined formats for exchange messages, and each message has an exact significance intend to provoke a defined response of the receiver.

The architecture of monitoring, control, and communications of the smart meter in distribution network predates the many advances made in the last 30 years in the fields of computing, networking, and telecommunications [10]. It has been observed that the speed of development and widespread adoption of smart meter networks will depend on same communication protocol or national standards that are applicable in a number of major areas [11, 14] which are as follows:

End-to-end cyber-security for smart meter networks: The broad communications infrastructure deployed for smart network applications may increase the potential number of attack points on the electricity network increasing the risk posed from cyber-security on energy supply. The importance of end-to-end cyber-security protection has been established to address the security concerns of smart networks. The technology choices deployed for smart networks will heavily depend on the setting of cyber-security policies.

Interconnection and interoperability standards for energy connection: The connection of widespread micro-generation solutions is anticipated to increase significantly with the adoption of smart networks. These new energy sources are predominantly focused on renewable technologies and low carbon emitting alternative fuel sources. While there are existing standards already available for connecting new loads, it is suggested that these standards need to be reviewed and modified to cater for interconnection of devices onto the network and this interoperability standards should be developed to use in the broader smart meter network rollout in any region of Australia.

Application-level data communications standards Application-level data communications standards are required to establish in distribution system that will enable interoperability and technology advancement to facilitate the national smart networking. These standards will be needed to enable a device in the field to converse with back-office systems using a common protocol that is interoperable. It is suggested that these standards should be leveraged from existing local and international communications standards.

5.3.1 Communication Protocols

Information has become a vital factor to the efficient operation and expansion of a reformation electric utility industry. Efficient communication with redundancy is a vital issue in any network that communicates information among nodes or stations [15]. For efficient and effective power system communication, TCP/IP, DNP3 protocol for data transmission is used widely. These protocols plays significant role in reliably transmitting data from different IEDs to main control center or server. Some of the most common protocols that are directly related to smart meter network are discussed in the following sections.

TCP/IP Internet Protocol Suite

TCP/IP provides connectivity between equipment from many vendors over a wide variety of networking technologies. It consists of a well-defined set of communication protocols and a several standardized application supports. TCP/IP communication protocols were originally developed for the Department of Defence (DoD) advanced research project agency's network (ARPANET), in the early 1970s. In the early 1980s, TCP/IP was included as an integral part of Berkeley's UNIX version 4.2. As a result, the protocol suit gained wide-spread network use. In 1983, TCP/IP become the military's protocol standard for networking and internetworking [16]. It is the protocol in use on both ARPANET and MILNET (a spin–off of ARPANET). The Internet Protocol (IP) is a network layer protocol which provides the routing function across multiple networks [16]. IP uses datagram to communicate over a packet-switched network [17, 18]. It also provides datagram services for Transmission Control Protocol (TCP) and User Datagram Protocol (UDP). This protocol comprises four layers:

Network Interface Layer: This layer is the lowest level of the TCP/IP stack. It is responsible for transmitting datagram's over the physical medium to their final destination.

Internet Layer: This layer is responsible for providing host to host communication. It is here that packet is encapsulated into an Internet diagram, the routing algorithm is run, and the datagram is passed to the Network Interface Layer for transmission on the attached network.

Transport Layer: This layer is responsible for providing communication between applications residing in different hosts. By placing identifying information in the datagram (such as socket information), the transport layer allows process-to-process communication. Depending on the needs of the requesting application, the Transport layer of the TCP/IP stack may provide either a reliable service (TCP) or an unreliable service (UDP). In a reliable delivery service, the receipt of a datagram is acknowledged by the destination station.

Application Layer: This layer is the highest level of the Internet model. A few of the applications include Telnet, File Transfer Protocol, and Simple Mail Transfer Protocol. Functional layers of TCP/IP and Open System Interconnection (OSI) layer are given below as shown in Fig. 5.4.

The IP forms a computer network by linking computers assigning each one a single IP address [19]. Each IP packet carries an IP address [20], which consists of destination address and host address. The host address is the IP address of the

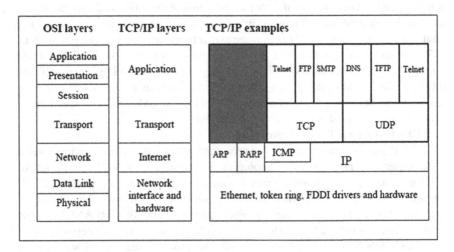

OSI layers	TCP/IP layers	TCP/IP examples						
Application	Application		Telnet	FTP	SMTP	DNS	TFTP	Telnet
Presentation								
Session								
Transport	Transport				TCP		UDP	
Network	Internet	ARP	RARP	ICMP		IP		
Data Link	Network interface and hardware	Ethernet, token ring, FDDI drivers and hardware						
Physical								

Fig. 5.4 TCP/IP protocols and functional layers [45]

sending computer, and the destination address is the address of the recipient or recipients of the packet. This protocol has also routing facility which route data from one network to another network from source to the destination address.

The major concern with IP is that it makes no attempts to decide if packets get to their destination or to obtain corrective action if they do not. Therefore, IP does not present guaranteed delivery. This difficulty can be avoided in some applications where a transport protocol that carries out such a function is utilized. The best example for the later is TCP [21], which makes up for IP's deficiencies by offering reliable, stream-oriented connections that are independent of the nature of the applications. However, other applications requiring best-effort services (faster transmission times) usually use UDP [22], which is a simple connection less transport layer protocol without guaranteed for reliable delivery. UDP packets are delivered the same as the IP packets and may even be discarded before reaching their destinations. Although the transmission of data requires the best-effort service in distribution network applications, reliability is also a major concern. The best-effort service requires the use of IP alongside TCP, which is the reliable transmission for distribution network. The highest level protocols within the TCP/IP protocol stack are application protocols. They communicate with applications on other hosts and are the client-visible interface to the TCP/IP protocol suite [23]. The TCP/IP protocol suite includes application protocols, namely:

• File Transfer Protocol (FTP);
• Simple Mail Transfer Protocol (SMTP) as an Internet mailing system.

These are a number of the most broadly implemented application protocols, but many others exist. Each particular TCP/IP implementation will contain a lesser or greater set of application protocols. It is often easier to make applications on top of TCP because it is a reliable stream, connection-oriented, congestion-friendly, and flow control-enabled protocol. As a result, most application protocols will use TCP and most applications utilize the client/server model of interface as well [23]. TCP is a peer-to-peer, connection-oriented protocol. The applications typically use a client/server model for communications as demonstrated in Fig. 5.5.

Fig. 5.5 The client/server model of applications [23]

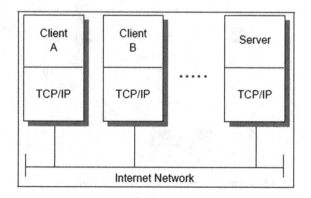

The server offers a service to clients when a client requests a service. The applications consist of both server and client, which can run on the same or on different systems. Users usually invoke the client part of the application, which builds a request for a particular service and sends it to the server part of the application using TCP/IP as shown in Fig. 5.6. The server receives a request, performs the service, and sends back the result in a reply. A server can usually manage multiple requests and multiple requesting clients at the same time [23].

If two hosts on different networks wish to communicate, the source host sends the packets to the right router. The router then routes each packet through the system of routers and networks until it reaches a router linked to the network as the destination host. Routing architecture is given in Fig. 5.7. This final router forwards the packet to the physical address of the destination host. Each network acts as a link between a router and all the hosts residing on it.

A router looks like a typical host to any of its connected networks. Routers forwards packets based on the destination network number rather than the physical address of the destination host. Since routing is based on numbers of network, the amount of information a router requirements is proportional to the number of networks that make up the Internet, not the number of machines.

In this study, TCP/IP is used in the OPNET simulation software environment as a communication protocol to develop the smart meter network in distribution system. In this model of smart meter network, latency or propagation delay is measured by selecting different bandwidth to evaluate the performances of smart meter distribution network.

Distributed Network Protocol (DNP3)

DNP3 [24] is a supervisory control and data acquisition (SCADA) protocol that permits information to be sent between a slave device (such as IED or smart meter) and a master device (such as control center or server). The slave device will

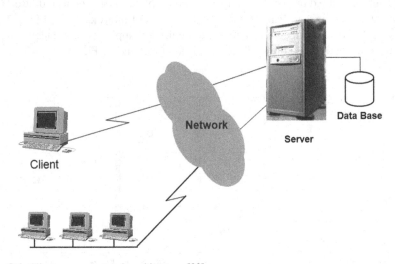

Fig. 5.6 Client–server network architecture [23]

Fig. 5.7 Routing
architecture [46]

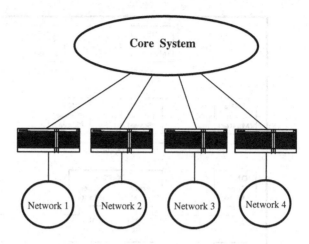

respond to requests for data that are issued by the master, but may also be configured to send data in respond to a field event without that data having been requested by the master. It has been used primarily by electric utilities like the electric companies, but it operates suitably in other areas.

Figure 5.8 shows common system architecture utilize today. At the top row of the figure is a simple one-to-one system having one master station and one slave. The second type of system is known as a multi-drop design. One master station communicates with couple of slave devices. The master desires data from the first slave, then moves onto the next slave for its data and repeatedly interrogates each slave in a round-robin order. The middle row illustrates hierarchical type system where the device in the middle is a server to the client at the left and is a client with respect to the server on the right.

Both lines at the bottom of Fig. 5.8 illustrate that data concentrator applications and protocol converters. A device may collect data from multiple servers on the right side of the figure and store this data in its database where it is recoverable by a master station client on the left side of the figure. This design is often shown in substations where the data concentrator gathers information from local intelligent devices for transmission to the master station. In current years, some vendors have utilized TCP/IP to convey DNP3 messages. This approach has allowed DNP3 to acquire advantage of Internet Technology and allowed economical data collection and control between widely separated devices.

The DNP3 software is layered to give reliable data communication, and also, this layering gives an organized approach to the communication of data and commands. Figure 5.9 shows the layering.

The link layer is responsible for making the physical link reliability. It performs this by providing error detection and duplicate frame detection. The link layer sends and receives packets, and the packets are called frames in DNP3 terminology. Sometimes transmission of more than one frame is essential to transport all of the information from one device to another. It is the task of the transport layer to break long messages into smaller frames sized for the link layer to transmit

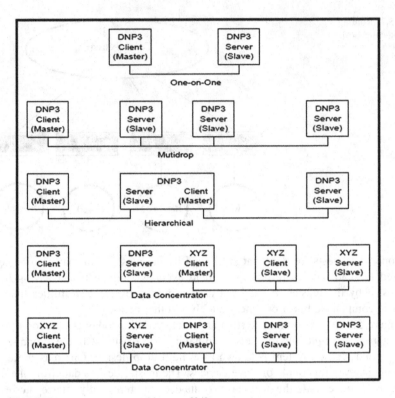

Fig. 5.8 DNP3 common system architecture [24]

or when receiving to reassemble frames into longer messages. In DNP3, the transport layer is incorporated into the application layer.

Application layer messages are broken into fragments. Fragment size is established by the size of the receiving device's buffer. It normally ranges from 2,048 to 4,096 bytes. If a message is larger than one fragment, it requires multiple fragments. Fragmenting messages has the responsibility of the application layer.

The application layer works together with the transport and link layers in order to allow reliable communications that offers standardized functions and data formatting. DNP3 goes a step further by classifying events into three classes. When DNP3 was conceived, class 1 events were regarded as having higher priority than class 2 events, and class 2 were higher than class 3 events. The user layer can request the application layer to poll for class 1, class 2 or class 3 events, or any combination of them.

The most attractive reasons for choosing the Internet protocol suite as a transport mechanism for DNP3 are as follows:

- Seamless integration of the substation LAN to the corporate WAN utility
- Leverage existing equipment and standard.

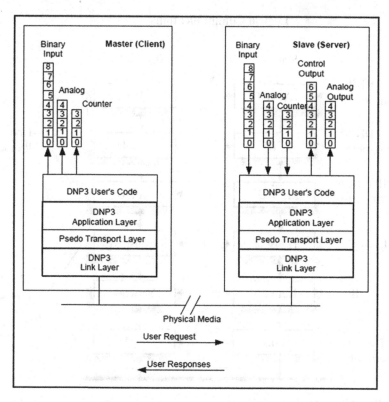

Fig. 5.9 Client and server relationship [24]

The Internet protocol suite and DNP use the OSI layering paradigm; each piece of the protocol stack in one station logically communicates with the corresponding piece in the other station(s). It is therefore easy to build DNP on top of the Internet protocol suite since the Internet layers appear transparent to the DNP layers as shown in Fig. 5.10.

5.3.2 Communication Standard

The increased expansion of the Internet and networked communications and the large-scale installation of wide-area networking technology facilitate the use of smart meter in distribution network [25]. A potential barrier is the threat related to installation of those new technologies without agreed standards. Lack of standards increase the threat of stranded assets; for example; utility deploying a tools is not supported by the industry and necessitating further setting up of some applications prior to the end of their expected lifetime. Standards can also assist to decrease setting up complication, make possible interoperability, and address security.

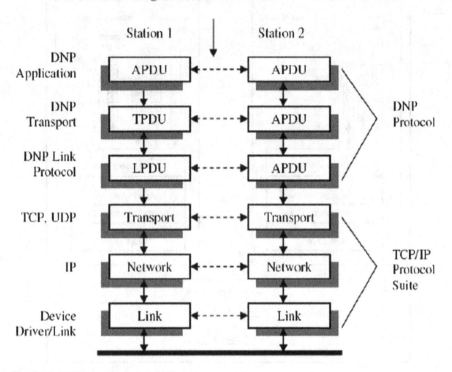

Fig. 5.10 DNP3 protocol stack [24]

Interoperability can offer appliance producers with the confidence and inspiration to install smart meter network in the products [26].

The necessity of smart meter system interoperability for the customer is already well recognized within the industry. Along with the challenge of developing the three different sets of standards (substation, feeders, and customer devices), comes the coordination between the different standards. The development of the smart meter network interoperability standards is a dynamic process which will have to adapt to the new type of loads and generation systems (DER and DG) to be connected to the distribution system [27].

Figure 5.11 shows an integrated distribution system based on interoperability standards such as the International Electro Technical Commission (IEC) 61850 series leading to a "plug and play" concept for the distribution equipment to be connected with the future Smart Grid. The illustrated new communications infrastructure (based on interoperability standards such as the IEC 61850 series) must have sufficient bandwidth to support the traditional distribution network as well as bulk data download to support advanced waveform and sequence of events analysis. The new infrastructure must be robust enough to support increased bandwidth requirements of IEC 61850 without degrading the existing communications with the automated distribution network [27]. Details of IEC 61850 and its use in smart meter network are discussed in the following sections.

IEC 61850

Development of IEC 61850 is based on the standard communication protocols to allow interoperability of IEDs from different manufacturers. IEC 61850 describes utilizing of existing standards and commonly accepted communication principles, which permits for the free exchange of information between IEDs. It allows applications to be designed independent from the communication theory enabling them to communicate using different communication protocols. Therefore, it provides a neutral interface between application objects and their related application services as shown in Fig. 5.12.

One of the most important features of IEC 61850 is that it covers qualitative properties of engineering tools, measures for quality management and configuration management. This is needed when utilities are planning to build a substation automation system (SAS) with the intention of merging IEDs from different manufacturers; they expect not only interoperability of functions and devices, but also a homogenous system handling [28]. Hence, IEC 61850 offers a neutral interface between application objects and the related application services allowing a compatible exchange of data among the components of a SAS [29, 30]. The IEC61850 abstract communication service interface (ACSI) models are conceptual

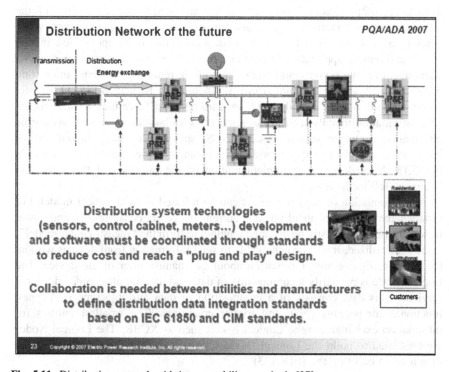

Fig. 5.11 Distribution network with interoperability standards [27]

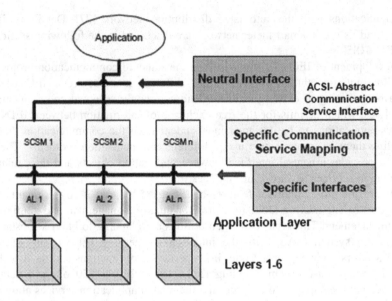

Fig. 5.12 The basic reference model [28]

definitions of common utility communication functions in field devices mainly describing communications between clients and remote servers. Some of the interfaces depict communications between a client and a remote server, while other interfaces are offered for communication between an application in one device and remote application in another device. The communications between a client and a remote server could be device control, reporting of events and setting control [31]. It aims for common utility functions to be performed consistently across all field devices provided [32]. Accordingly, ACSI defines substation-specific information models such as common DATA classes and substation-specific information exchange service models. ACSI specifies the basic layout for the information models and the information exchange service models as shown in Fig. 5.13. Nevertheless, the implementation of the objects and the modelling issues are left to the user.

A representation of physical object can be referred to as an object model. For instance, the measurements of voltage, current and power consumption, and power factor in a relay can easily be grouped together to form the "measurement model". Once standardized, it will be achievable to send information from devices without having to recognize any information about the manufacturer of the device. The Logical Node is primarily a composition of data and data set plus some additional services. This data consists of a composition of data attribute type (DAType), functional components (FC), and trigger conditions. The smallest entities for information exchange are the Logical Nodes such as XCBR. The Logical Nodes are then used to build the Logical Devices. In turn, several Logical Devices are then used to build up the IEDs [33].

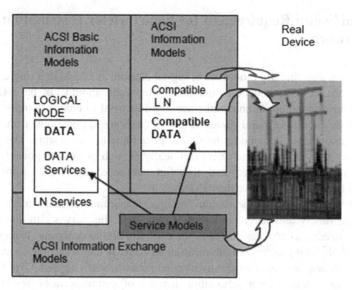

Fig. 5.13 ACSI conceptual model [32]

Each of the classes comprising the Logical Node consists of a number of building blocks. Even though ACSI permits separate devices to share data and services, it is only an abstract application layer protocol without any real procedure for sending and receiving data. It can only be serviceable when it is plotted to a definite communication service such as Manufacturing Message Specification (MMS) protocol, Distributed Component Object Model (DCOM) or Common Object Request Broker Architecture (CORBA). The specific communication service mapping (SCSM) explains the execution details of services and models using a specific communication stack [34]. The new communications infrastructure (based on interoperability standards such as the IEC 61850 series) must have sufficient bandwidth to support the traditional distribution network as well as bulk data download to support advanced waveform and sequence of events analysis. The new infrastructure must be robust enough to support increased bandwidth requirements of IEC 61850 without degrading the existing communications with the automated distribution network [35].

Integration of large-scale smart meter in distribution network introduces new challenges in the network as considerable amount of bandwidth is required for data communication between smart meter and central station. For the distribution network, it is an essential need to consider propagation delay and latency. In the next few sections, major criteria for a robust smart meter network are discussed.

5.4 Bandwidth Requirement for Smart Meter Distribution Network

In the digital age, literally thousands of digital data are available in a single IED in smart meter network, and communication bandwidth should not be a limiting factor. A broadband communications network is a need to enable comprehensive system-wide monitoring and coordination to facilitate applications such as distribution automation, demand response, and power quality monitoring. The present SCADA systems do not sense or control nearly enough of the components of the modern distribution system, and therefore, reliable, up-to-date information embedded power system is required [36].

Smart meter network needs an intelligent communications infrastructure facilitating timely and secure information flow, and this network in distribution system is needed to provide power to the evolving digital economy. It must offer robust, reliable, and secure communication since IEDs and smart meter make the necessary system assessments for distribution network when needed. Over the past 35 years, there has been a substantial increase of communication speeds from 300 bps (bits per second) to digital relays that today operate at 100 Mbps. Not only have the communication speeds changed, but the communication protocols have migrated from register-based solutions to text-based data object requests. In addition, the physical interfaces have transitioned from RS-232 serial over copper to Ethernet over fiber—both local and wide-area network. Interoperability has become a reality, and today's devices are self-describing and programmable [37]. In this digital distribution system, a utility pole hooked up to high-speed Internet. Hundreds of IEDs attached to the lines monitor how power flows through the consumer's house and that information is then sent back to the utility company. The process allows a utility to more efficiently handle the distribution of electricity by allowing two-way communications between consumers and energy suppliers via the broadband network [38]. Smart meter network also requires a far greater degree of visibility and control devices as data collection and real-time analysis become a more fundamental part. However, real-time measurements with communication channels are very limited in current distribution networks. Moreover, distribution network is enhanced by large-scale application of smart meters [39, 40]. Smart meter network allows easy broadcast of network applications from center controller to many connected smart meters but only slow communications back from the connected devices to the central controller [41].

An inspection of the Victorian automated metering infrastructure (AMI) functionality specification [42] shows a system that reveals a change in capability relative to the pre-existing meter infrastructure. This functionality of AMI is designed to deliver specific outcomes. Some more advanced smart grid concepts will lay beyond the capabilities of an AMI with a reasonable level of sophistication. Key restrictions are the asymmetric bandwidth of the communications

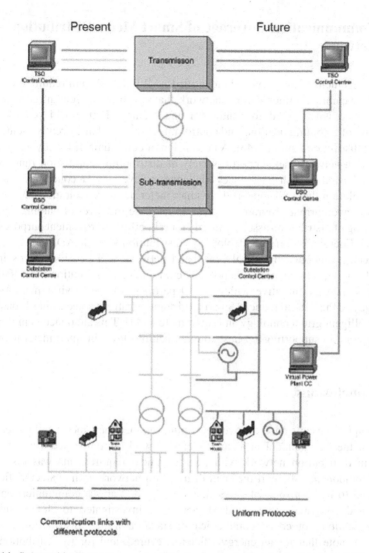

Fig. 5.14 Smart grid rollout of communications coverage to the distribution level [42]

channel and latency. Load control is one of the faster AMI commands. The Victorian functional specification [42] requires 99 % of meters respond in 1 min to group commands but for individual meter commands only 90 % need respond in 30 min. Only 2 % of meters may be switched individually within a 24 h period. Distributed generation, storage, and load control would be unable to operate within the limitations of an AMI. Any communications to individual devices are potentially subject to very long delays [41].

5.5 Communication Coverage of Smart Meter Distribution Network

The management of the previously identified AMI functions will require the use of a highly capable communications network that can provide guaranteed levels of performance with regard to bandwidth and latency. Figure 5.14 provides the extension of a communications and control infrastructure that currently include the existing distribution system [43]. An extension of communications coverage to the distribution network can support a variety of distribution automation functions.

Each Australian distributor faces challenge in developing a coherent communications solution that will support their smart meter network aspirations. Many distributors have current programs in place to increase the level of automation, and monitoring of distribution assets primarily for reliability improvement purposes. For example, Energex is currently deploying a distribution level SCADA network [41]. The recently announced federal broadband network will provide speeds in the 100 Mb/s range. It may present a possible communications solution [41]. National Energy Efficiency Initiative within the Department of the Environment, Water, Heritage and the Arts announced Smart Grid demonstration projects links broadband with intelligent grid technology and smart meters [44]. This announcement marks a turning point in supporting a robust communications layer in smart meter network.

5.6 Conclusions

This chapter presents an extensive description on the smart meter and its requirement for the development of a smart meter network. Initially, necessity of smart meter in distribution network with their technical requirements was discussed. Later, components of smart meter in distribution network with characteristics are discussed to make a foundation for the smart grid. Various communication protocols and communication standard were also investigated to choose suitable communication protocols or communication standard in smart distribution network.

It is to note that for an energy-efficient, secured, and robust distribution network, integration of smart meter with distribution network is essential. As discussed, communication protocols and communication standard and bandwidth are vital factors for an effective smart meter network.

References

1. Marvin S, Chappells H, Guy S (1999) Pathways of smart metering development: shaping environmental innovation. Comput Env Urban Syst 23(2):109–126
2. Energy retail association, smart meters, [Online]. Available at. http://www.energy-retail.org.uk/smartmeters.html

3. Gerwen RV, Jaarsma S, Rob Wilhite K (2006) The Netherlands, smart metering
4. Technical paper, reliable and precise overview over electrical characteristics and power. Available at. http://www.siemens.com.au
5. Forrest J, smart metering Technical report: capgemini, consulting, technology, outsourcing. Available at. http://www.uk.capgemini.com/industries/utilities/smart-metering/
6. Strbac G (2008) Demand side management: benefits and challenges. Energy Policy J 36(12):4419–4426
7. Smart metering with a focus on electricity regulation (2007) Technical report on European regulator's group for electricity and gas
8. Hendricks G (2009) Wired for progress-building a national clean-energy smart grid technical report: center for American progress, Feb 2009. Available at. http://www.americanprogress.org/issues/2009/02/pdf/electricity_grid.pdf
9. Halliday C, Urquhart MD, Network monitoring and smart meters. Available at. http://elect.com.au/Attachments/Network%20Monitoring%20and%20Smart%20Meters.pdf
10. Smart grid, wikipedia foundation Inc. Available at. http://en.wikipedia.org/wiki/Smart_grid as at 25th July 2009
11. Smart networks position paper, technical report: energy network associations, Australia, Sept 2009. Available at. http://www.ena.asn.au/udocs/2009/09/Smart-networks-policy-paper-2.pdf
12. Heydt GT, Kezunovic M, Sauer PW, Bose A, MCalley JD, Singh C, Jewell WT, Ray DJ, Vittal VV (2009) Professional resources to implement the smart grid, technical report: north american power symposium. Available at. http://www.pserc.wisc.edu/ecow/get/publicatio/2009public/heydt_professional_resources_smart_grid_2009_adobe7.pdf
13. Collier SE (2009) Ten steps to a smarter grid, IEEE 2009 rural electric power conference. Available at. http://www.milsoft.com/downloads/presentations/10%20Steps%20to%20a%20Smarter%20Grid%2001302009.pdf
14. Smart grid, smart city, summary of stakeholder workshop, technical report: the national energy efficiency initiative. Available at. http://www.environment.gov.au/smartgrid/pubs/smartgrid-workshop-summary.pdf
15. Ericsson GN, Torkilseng A (2005) Management of information security for an electric power utility ¾ on security domains and use of ISO/IEC17799 standard. IEEE Trans Power Deliv 20(2):683–690
16. Stallings W (2004) Computer networking with international protocols and technology, Pearson education, Inc
17. Forouzan BA (2003) TCP/IP protocol suite, ISBN: 0072460601, 2nd edn. McGraw-Hill Professional, New York, pp 19–47
18. Wright GR, Stevens WR (1995) The protocols: TCP/IP illustrated, ISBN 020163354X, vol 1. Addison-Wesley Professional, Boston, pp 33–53
19. Fairhurst G (2005) The internet protocol (IP), web doc., department of electrical engineering, University of Aberdeen, UK, 2001, viewed June 2005. [Online]. Available. http://www.erg.abdn.ac.uk/users/gorry/course/inet-pages/ip.html
20. Apple Computer Inc (2005) Inside macintosh: networking with open transport/part 1—open transport essentials: chapter 11—TCP/IP services, Version 1.3, Web Doc., 15 Jan 1998, viewed Jan 2005. [Online]. Available. http://developer.apple.com/documentation/mac/NetworkingOT/NetworkingWOT-52.ht ml
21. Comer DE (1988) Internetworking with TCP/IP: principles, protocols, and architecture, vol 1, 4th edn. Prentice Hall, Englewood Cliffs
22. Postel J (1980) User datagram protocol, RFC-768, USC/Information Sciences Institute, Aug 1980
23. International technical support organization TCP/IP tutorial and technical overview. Available at. http://www.redbooks.ibm.com/redbooks/pdfs/gg243376.pdf
24. DNP3 users group, —DNP3 specification, vol 8. IP networking, draft H, Dec 04. Available at. http://www.dnp.org/ftp/spec-ipnetworking/td-ipnetworking-draft-h.pdf

25. Networking the smart grid (2009) Technical report: a Tropos networks white paper, April 2009. Available at. http://www.smartgridnews.com/artman/uploads/1/NetworkingSmartGrid WP_A7.pdf
26. Smart grid, smart city, grand guideline (2009) Technical report: the national energy efficiency initiative, 30 Sep 2009. Available at: http://www.environment.gov.au/smartgrid/publications/pubs/smartgrid-grant-guidelines.pdf
27. Simard G, Clark L, Uluski B (2009) Nist interim smart grid standards interoperability roadmap workshop, technical report: contribution from IEEE PES distribution automation working group, May 2009. Available at. http://grouper.ieee.org/groups/td/dist/da/NIST%20INTERIM%20SMART%20GRID%20WORKSHOP%20IEEE%20PES%20DAWG%20CONTRIBUTIONMay1st09.pdf
28. Janssen MC, Koreman CGA, Substation components plug and play instead of plug and pray: the impact of IEC 61850, Kema T&D power, Netherlands
29. IEC 61850 website, IEC 61850 communication networks and systems in substations, May 05. [Available Online]. http://www.61850.com/
30. ObjectWeb consortium website, what is middleware, web doc. Available at. http://66.102.7.104/search?q=cache:pfkohQJSB7oJ:mIddleware.objectweb.org/+%22middleware+is%22&hl=en
31. Hammer E, Sivertsen E (2008) Analysis and implementation of IEC 61850 standard, Master's thesis
32. Mahmoud QH (2004) Middleware for communications, ISBN: 0470862068, Ed. Wiley, England
33. Schwarz K, Seamless real-time information integration across the utility enterprise to reduce costs, presented at the PowerGen Asia, Schwarz consulting company, SCC, Karlsruhe, Germany. Available at. http://nettedautomation.com/download/PowerGenAsia_2000_06_20.PDF
34. Ozansoy C, Zayegh A, Kalam A, Modelling of a network data delivery service middleware for substation communication systems using OPNET. In: Proceedings of the AUPEC'03 conference, Christchurch, New Zealand, 28 Sept–1 Oct, Paper no: 91
35. Simard R, Clark L, Uluski B (2009) Nist interim smart grid standards interoperability roadmap workshop, technical report: contribution from IEEE PES distribution automation working group, May 2009. Available at. http://grouper.ieee.org/groups/td/dist/da/NIST%20INTERIM%20SMART%20GRID%20WORKSHOP%20IEEE%20PES%20DAWG%20CONTRIBUTIONMay1st09.pdf
36. Networking the smart grid (2009) Technical report: a troops networks white paper. Available at. http://www.smartgridnews.com/artman/uploads/1/NetworkingSmartGridWP_A7.pdf
37. Sollecito L Smart grid: the road ahead, technical report: intelligrid architecture report: intelligrid user guidelines and recommendations, vol 1. EPRI. Available at. http://www.gedigitalenergy.com/multilin/journals/issues/Spring09/Smart_Grid_The_Road_Ahead.pdf
38. Hart K (2009) Engineering a smart grid for energy's future, technical report: the Washington post, April 2009, Available at. http://www.washingtonpost.com/wp-dyn/.../AR200904 2602628.html
39. Wu J, Jenkins N, Self-adaptive and robust method for distribution network load and state estimation
40. He Y, Jenkins N, Wu J, Eltayed M, ICT infrastructure for smart distribution networks
41. Wolfs P, Islam S, Potential barriers to smart grid technology in Australia. Available at. http://ieeexplore.ieee.org/stamp/stamp.jsp?tp=&arnumber=5356623
42. Department of primary industries, victoria, advanced metering infrastructure—minimum AMI functionality specification (Victoria). Available at. http://share.nemmco.com.au/smartmetering/Document%20library/Smart%20meter%20background%20info/Background%20-%20Minimum%20AMI%20Functionality%20Specification%20Vic%20-%20Sep%202008.pdf

43. European technology platform smartgrids, strategic research agenda for europe's electricity networks of the future, European commission, directorate general for research, directorate energy
44. Australian government, department of the environment, water, heritage and the arts, environment budget overview 2009–2010. Available at. http://www.environment.gov.au/about/publications/budget/2009/ebo/pubs/budget-overview-09-10.pdf
45. Apostolov AP (2002) Application of high-speed peer-to-peer communications for protection and control, ALSTOM T&D protection and control, CA, [Online]. Available. http://www.tde.alstom.com/p-c/ftp/docs/papers/CIG RE34AA.pdf
46. Technical paper, TCP/IP protocol. Available at. Internet

Chapter 6
Demand Forecasting in Smart Grid

A. B. M. Shawkat Ali and Salahuddin Azad

Abstract Changes in temperature, rainfall, icefall, sea level, and the frequency and severity of extreme events are raising a question that how much energy we should produce to meet the world demand. The smart grid is a new paradigm that enables two-way communications between the electricity providers and consumers. Smart grid emerged due to the initiatives by the engineers to make the power grid more stable, reliable, efficient, and secure. The smart grid creates the opportunity for the electricity consumers to play a bigger role in their power usage and motivates them to use power sensibly and efficiently. Hence, in the implementation of smart grid, demand management going to play a vital role. Demand scheduling is an effective way to implement demand management at the customer side. It is an automated and intelligent method to shift a portion of the demand from peak to off peak so that the demand curve is flattened. To optimize the demand scheduling, the accurate energy usage pattern of the consumers is essential. This is where the demand forecasting comes into play. This chapter investigates how effectively the machine learning algorithms can forecast the electricity demand to facilitate electricity demand management. For the experiments, a real-life dataset is considered which was collected locally at Rockhampton, Australia. From the experimental experience, it is concluded that support vector machine is the most reliable machine learning tool for accurate prediction of electricity demand.

6.1 Introduction

With the growth of world population, the demand for electricity is rapidly increasing. The principal source of electricity generation is fossil fuels. In 2009, 67 % of the electricity produced worldwide was from fossil fuels, while 16 % of

A. B. M. S. Ali (✉) · S. Azad
Central Queensland University, Rockhampton QLD 4702, Australia
e-mail: s.ali@cqu.edu.au

A. B. M. S. Ali (ed.), *Smart Grids*, Green Energy and Technology,
DOI: 10.1007/978-1-4471-5210-1_6, © Springer-Verlag London 2013

the electricity was from renewable energy sources such as solar, air, biomass [1]. Burning coal, oil, and natural gas release CO_2 into the atmosphere. As a consequence, the worldwide CO_2 emission reached an upsettingly high historical maximum of 30,600 million of tons in 2010, according to the International Energy Agency (IEA) [2]. The CO_2 emission is the most important cause of warming the earth and changing the climate in other ways. For instance, it may change the severity and duration of storms or droughts. The climate change can eventually affect heating, cooling, water use of the daily life as well as the sea level. In the wealthy countries, the average cost of climate change would probably be small, although some people and regions might have high costs and others might receive large benefits. Conversely, in some poor countries, the cost could be very high as they are susceptible to drought, water shortages, and crop failure. In addition, a large or fast change in climate might have a big effect on plants and animals in the natural environment. A very rapid climate change is unlikely, but could be catastrophic, even for wealthy countries [3]. Consequently, there is a strong push to resist the climate change to minimize its consequences.

Electricity demand refers to the amount of electricity is needed by households, businesses, and industries in a country. In general, the demand for electricity can be more than double the average demand on a typical day, which is called *peak demand*. Past experiences show that peak demand may rise excessively a few times each year on extremely hot summer days when the air conditioners are in operation in households in addition to other appliances, while the commercial and industrial sector is consuming power. In the same way, in the extremely cold weather, the heating systems cause a peak demand of electricity. In both cases, when the demand exceeds the generation capacity, it can lead to power outages and catastrophic failures in the grid. The electricity *demand management* is a mechanism to motivate the electricity consumers to use less energy during the peak time so as to reduce the investment on electricity production and ensure a reliable supply. The typical measures for demand management are as follows [4]:

- Pricing changes—The customers receive incentive for reducing their electricity usage during peak times. Alternately, the price of the electricity is made cheaper during the off peak.
- Direct load control—It is a load shedding mechanism managed where the utility reduces the energy consumption of home appliances by controlling its operation. The customers are offered incentives for helping relieve the local peak by cutting energy usage.
- Giving customers a better idea of their electricity usage. Smart meter offers in-home displays, so that consumers are able to follow their electricity consumption and have greater control over their electricity consumption during peak times.
- Promoting distributed generation solutions where electricity is produced near the customer premises, such as solar panels or wind turbines to minimize the local peak demand.

- Encouraging electricity customers to set up a power storage system along with their own solar panels which could meet the additional demand during the peak to avoid blackout in the grid.

Load scheduling [5, 6] is an effective way to implement demand management at the customer side. It is an automated and intelligent method to shift a portion of the demand from peak to off peak, so that the demand curve is flattened. The load scheduling is driven by the change in price and the incentive paid by the utilities and minimizes customers utility bill or maximizes the incentive receive from the utility. Hence, pricing is an important tool to shape the demand curve. The price of electricity during the peak is higher than the off peak, so that the consumers tend to reduce the demand during the peak. By the fixing the price optimally, the utility can effectively flatten the demand throughout the day. The load scheduling achieves substantial saving in peak demand as well as overall operational cost and carbon emissions. To optimize load scheduling, the accurate energy usage pattern of the consumers must be known in advance. This is where the demand forecasting comes into play.

The organization of this chapter is as follows. Section 6.2 addresses the existing literature on the demand forecasting methods which already gained popularity among the researchers. The top most three popular forecasting algorithms are summarized in Sect. 6.3. An experimental demonstration with proper outcome analysis is figured out in Sect. 6.4 Finally, the conclusion of this chapter with the future directions is noted toward the end of this chapter in Sect. 6.5.

6.2 State-of-the-Art

The operation principles and the components of the existing electrical power grid have entered into a new era which is called *smart grid*. Among the significant objectives of the smart grid, the demand management is one which plays a key role in increasing the efficiency of the grid [7]. Demand management enables the consumption of electricity in a smart way, so that the investment on generation and distribution is minimized. An effective demand management scheme requires accurate prediction of demand and price, based on the knowledge from the historic data.

In reality, the load and price forecasting is widely studied in the literature. The literature on load forecasting extends as far back as the mid-1960s [8, 9]. In general, load forecasting is essential for dispatchers, who are the commercial or governmental bodies responsible for dispatching electricity to the grid [10]. Load forecasting provides tools for operation and planning of a dispatcher where decisions such as purchasing or generating power, bringing peaker plants online, load switching, and infrastructure development can be made [11]. Electricity market regulators and dispatchers rely on forecasting tools that provide short-, medium-, and long-term forecasts. Short-, medium-, and long-term forecasting of demand management are necessary to operate a power grid in a smart way.

Short-term demand forecasts cover hourly or daily forecasts, medium-term forecasts consider a week to a month, and long-term forecasts cover several months to a year. Based on the terms, forecasting techniques may also differ according to the time interval [12]. The automated load scheduling scheme [13] aims to minimize the consumer expenses as well as the waiting times of the delayed demands. The scheduling scheme is augmented with a price predictor in order to anticipate the prices of the next few hours ahead. The prediction is essential in the situation when the grid operator only announces the prices for the next one or two hour.

In demand management, regression is a widely used statistical technique for load forecasting. In general, regression methods aim to model the relationship between the load and the environmental factors, for example, temperature, rain, cloud [14]. Regression considers time series data to achieve the appropriate forecasting in the load management area. As a result, the alternative name of regression is called time series methods. In the literature, a wide variety of time series methods such as autoregressive moving average (ARMA), autoregressive integrated moving average (ARIMA), autoregressive moving average with exogenous variables (ARMAX), and autoregressive integrated moving average with exogenous variables (ARIMAX) methods are also employed to do load forecasting [10]. Also, in the extended literature, some other tools including various types of Box–Jenkins time series approaches [15], adaptive forecasting techniques [16], co-integration analysis [17], seasonal integrated autoregressive moving average models (SARIMA) [18], two-level seasonal autoregressive models (TLSAR) [19, 20], autoregressive fractional integration moving average models (ARFIMA) [18], dummy-adjusted seasonal integrated autoregressive moving average models (DASARIMA) [19, 20], smooth transition periodic autoregressive models (STPAR) [21] are also used to perform the load demand forecasting. The latest literature also shows that artificial neural network (ANN), support vector machine (SVM), linear regression, and fuzzy logic are also very popular forecasting techniques.

Researchers in the power engineering domain have been dedicated to modeling and forecasting the short-term electric load. Most of these methods in the literature can be classified into two broad categories: (1) *artificial intelligence* (AI) and (2) *statistical* approaches. In the AI-based approaches, artificial neural networks (ANNs) have received the most attention by both the academia and the industry [22, 23]. However, in the statistical approach, multiple linear regression is the most widely applied technique for short-term and mid-term electrical load forecasting [24–26].

Considering the other aspect of load modeling, the regression-based models have also been applied to a wide field in power systems such as load monitoring at distribution substations [27]. However, in the earlier times, when the regression methods started to be deployed, Hong et al. [28] believed that due to the quality of data sources and limited capability of computers, the power of regression methods could not be fully demonstrated in the utilities. On the other hand, modern techniques allow utilities to record and store historical data of load and weather in sufficiently high quality in terms of both resolution and correctness. Nowadays, modern computing environment allows engineers and researchers to perform computational intensive analysis over large-scale data. For example, "in the case of applying the regression analysis to the dataset containing 4 years of hourly load and temperature, it takes less than 5 min to update the parameters of the proposed model in a computer with a 3.5 GB RAM and 2.2 GHz CPU" [28]. To further improve the linear regression in the electrical load forecasting domain, researchers have proposed the fuzzy linear regression (FLR) technique. FLR problems can be divided into four categories based on whether the inputs and outputs are fuzzy or not [29]. Basically, there are two ways of solving an FLR problem: linear programming (LP) methods and fuzzy least squares methods [30]. The first one aims to minimize the average fuzziness [31] or to maximize the average membership [32] given that the fuzzy forecast contains the actual value to a certain degree. Redden et al. [33] compared various FLR methods and found out the differences between fuzzy and traditional regression approaches. Donoso et al. [34] pointed out the weakness of the proposed approach. For instance, the optimization of the central tendencies was not considered. Usually, a high number of crisp estimates are derived in fuzzy regression [34]. Bárdossy [35] used different fuzziness measures instead. The later way tries to minimize the sum of the square of the residuals, which is a quadratic solution [36]. Donoso et al. [34] constructed a quadratic non-possibilistic (QNP) model, which is the quadratic error for both the central tendencies, and each one of the spreads is minimized.

In parallel with the linear regression methods, nonlinear algorithms for instance, ANN, SVM have also been used to solve demand forecasting. There are two differences between a linear regression and an ANN model: (1) the regression model is linear in parameters and (2) there is no hidden layer functions as there are in ANN models [37]. The AI-based techniques have been proven to be a useful tool over a period; however, the main drawback of this model is that the training of the model can take a great deal of time [38]. The back-propagation (BP) learning algorithm [39] is an iterative gradient descent procedure. It is very capable of handling such large learning problems. In addition, some remain remarks are as to the performance of ANN and whether they truly outperform standard forecasting methods [40]. ANN is one of many computational methods that has already gained attention in research and vastly used in prediction application. It is a well-known fact that ANN can model any nonlinear relationship to an arbitrary degree of accuracy by tuning the network parameters. In addition, it can handle nonlinearities among variables as the expected nature of the energy consumption data is nonlinear [41]. It has been proved as an effective technique in the energy sectors such as optimization and generalization ability, adaptability, a legacy of information processing, failure tolerance, and low power consumption [42–45]. The authors present several applications of ANNs in energy problems, such as modeling and designing a solar steam generating plant, estimation of a parabolic-trough collector's intercept factor and local concentration ratio, modeling and performance prediction of solar water heating systems. There are also demand management strategies that have been designed using a host of methods and techniques including artificial neural networks amidst other soft computing techniques [46, 47]. There are different types of ANNs in the literature for demand management including BP [39], multilayer perceptron (MLP) [48], fuzzy NN [49], radial basis function (RBF) NN [50]. The hybrid NN is also a popular method for forecasting. Khotanzad et al. [39] built a combination of three networks—MLP trained with BP, MLP trained with Levenberg–Marquardt (LM) algorithm, and functional-link network. Lee et al. [51] proposed an ANN model to forecast the electricity loads of weekdays and weekends. This model does not yield less relative error than other approaches. Another study [52] proposed a 3-layer back-propagation ANN to deal with daily load forecasting issues. The inputs include three indices of temperature—average, peak, and lowest. The expected outputs are the peak loads. The model provides more accurate forecasting results than the regression model and the time series model. Among these ANN algorithms, back-propagation algorithm is the most popular algorithm over decades up until now.

Comparatively new but one of the effective algorithms is SVM [53]. As like demand management, it shows the effectiveness in other areas as well. A comparative study is conducted [54] between SVMs, ANNs, and the traditional seasonal autoregressive model (SAR) in the forecasting of lake water levels. They found that SVM is generally compatible with the other two models, but in the long-term forecasting, SVM shows a better performance. Another study in [55] applied SVMs for option pricing. Jayadeva and Chandra [56] successfully used the regularized least squares fuzzy support vector regression for financial time series forecasting, while Zhang et al. [57] used support vector regression for online health monitoring on a large-scale data. Another well-cited research [58] conducted a study to predict chaotic time series using SVMs. The SVMs performance stood out when compared to other approximation methods such as polynomial and rational approximation, local polynomial techniques, and ANNs. The research by Mohandes et al. [59] employed SVM for the prediction of wind speed. The research found the performance of SVM comparable to the established algorithm such as ANN. Another study [60] tested SVMs performance in forecasting financial time series, and it shows that SVM is superior to multi-layer back-propagation neural network in forecasting financial time series. A dynamic SVM model (DSVM) has been proposed in [61] to solve nonstationary time series problems. The experimental results demonstrate that the proposed algorithm outperforms the standard SVMs in forecasting nonstationary time series. In the same year, Tay and Cao [62] proposed C-ascending SVM to model nonstationary financial time series. The experimental results confirm that the C-ascending SVM with the actual ordered sample data consistently outperforms the standard SVM. Another study [63] applied SVMs to predict air quality. The experimental results demonstrate that SVM outperforms conventional radial basis function (RBF) network. Mohandes et al. [59] used SVM to predict the wind speed. The experimental results demonstrate that the SVM model outperforms the multi-layer perceptron neural networks in terms of root mean square errors. Zhou et al. [64] successfully used SVM for short-term load forecasting. Another research [65] combined principal component analysis (PCA) and SVM for long-term load forecasting. The generalization and learning ability of single kernel function SVM were weak in their study. The extension of this research [66] combined rough set theory and SVM for the same purpose. Empirical results indicate that SVM model provides more accurate results than the existing model, autoregressive integrated moving average (ARIMA) model, and general regression neural network (GRNN) model [67].

6.3 Electricity Demand Data

We have collected daily electricity demand data for a household in an hourly basis. In this data collection process, the loads for one refrigerator, one freezer, one electric stove, one microwave, one rice cooker, one toaster, five fans, five electric bulbs, one washing machine, one vacuum cleaner, three air conditioners, one phone set, one PC, and two laptops were taken into account. Also, a hot water system was considered, which mainly uses solar power. However, when the solar power is insufficient, it automatically switches to electricity. We split the solar power data from the total power consumption, which is used as the demand of hot water system in the forecasting model. The nature of data is numeric, and the length of data collection was 01/01/2011 to 06/30/2011.

6.4 Time Series Analysis

Analysis of time series has been a part of statistics from the beginning. Nowadays, it is widely applicable in the area of signal processing, pattern recognition, econometrics, mathematical finance, weather forecasting, earthquake prediction, electroencephalography, control engineering, communications engineering, and so forth. Time series analysis is based on the statistical theory and specially evolved methods and their applications on the time variant data. Data may be classified according to different criteria, and time is one of these criteria which may be used to estimate the future values. In general, data are categorized into three groups: (1) *cross section data*, (2) *time series data*, and (3) *pooled data*. Time series is a set of values of the given variable relating to one single entity of observation at different points in time which is called time series. Observations on a given attribute and their measured values are collected repeatedly at different points in time. However, the duration of the time period during the data collection remains the same. In general, the duration falls in an hour, day, week, month, one quarter of a year, year, even half or one decade. The time series window interval is always the same. Though the duration of time or gap between two periods remains constant, yet the time itself is a variable in time series. Time is the most important variable in analysis only in dynamic social, political, and economic systems; time is invariant in static or stationary systems. It is the system rather the time which is stationary in such systems. Time is stationary in the sense that the lapse or movement in time does not bring any change in the system. Analysis of time series data may perform the following functions, which make analysis of time series different from the analysis of cross section or pooled data [68, 69]. Some common characteristics of time series are as follows:

- Time series analysis may provide an explanation of relations and concepts that may underlie the data;
- It may constitute the base for budding certain concepts that may be associated with empirical generalizations;
- Time series analysis may disclose the nature and direction of inter-relations and changes therein, if any;
- It may be used to offer prediction of future values of the variable under consideration;
- Systematic trend or random behavior of data may be uncovered by the analysis of such data. Behavior of random factors and its outcome(s) involves high degree of risk and even volatility. Certain time series data, especially data relating to financial time series, are subjected to external shocks that impart great deal of volatility. These facets and factors they pertain to are not amenable to usual analytical methods and models.

Time series models can be classified into two groups. First, univariate models where the observations are those of single variable recorded sequentially over equally spaced time intervals. The other kind is multivariate, where the observations are of multiple variables. In this chapter, we consider the multivariate models: linear regression, multilayer perception, and SVM. The experimental outcome demonstrates that SVM is the most effective algorithm for load prediction and demand management.

6.5 Algorithm Description

Machine learning is originated from the root concept of Artificial Intelligence, and it is concerned with the development of techniques and methods that enable the computer to learn data. The data could be symbolic, numeric, image, and in many forms. In simple terms, development of algorithms which enable the machine to learn and perform tasks and activities is regarded as machine learning. Over the period of time, many techniques and methodologies have been developed for machine learning tasks, for instance multilayer perceptron, support vector machine, decision tree, and so on.

The concept of SVM was first published in 1992, introduced by Boser, Guyon, and Vapnik in COLT-92. SVMs are a set of related supervised learning methods used for classification and regression. They belonged to a family of generalized linear classifiers initially. Over the time, it has been extended to a nonlinear form. SVM is also a prediction tool that uses machine learning theory to maximize the prediction accuracy while automatically avoiding over-fit to the data. SVM can be defined as a system, which uses hypothesis space of a linear function in a high-dimensional feature space where data are very closely linearly separable, trained with a learning algorithm from optimization theory that implements a learning bias derived from statistical learning theory. The foundations of SVM have been developed by Vapnik [70] and gained popularity due to many promising features such as better experimental performance. Vapnik uses the structural risk minimization (SRM) principle in the SVM implementation, which has been shown to be superior to traditional empirical risk minimization (ERM) principle [71], used by conventional neural networks. SRM minimizes an upper bound on the expected risk, where as ERM minimizes the error on the training data. It is this difference which equips SVM with a greater ability to generalize, which is the goal in statistical learning theory. Initially, SVM was developed to solve the classification problem; later on, they have been extended to solve regression and clustering problems [72].

The main objective of forecasting for a given series $x_1, x_2, x_3, \ldots, x_N$ is to estimate future values such as x_{N+k}, where the integer k is called the lead time [73].

The forecast of x_{N+k} made at a time N for k steps ahead is denoted by $\hat{x}(N, k)$. Since SVM is a very similar function estimate-based algorithm, the function $f(x)$ is estimated rather than $\hat{x}$. In the following paragraph, the SVM regression (which will call SVR later) is summarized based on [74]. This technique is used for time series analysis later in this chapter.

The essence of the regression problem is to determine a function that can approximate future values accurately. The generic SVR estimating function takes the form:

$$f(x) = (w \cdot \Phi(x)) + b, \tag{6.1}$$

where $w \subset R^n$, $b \subset R$ and Φ denote a nonlinear transformation from R^n to high-dimensional space, theoretically up to infinite. Now, the next job is to find the value of w and b, such that the values of x can be determined by minimizing the regression risk:

$$R_{reg}(f) = C \sum_{i=0}^{\ell} \Gamma(f(x_i) - y_i) + \frac{1}{2} \|w\|^2, \tag{6.2}$$

where $\Gamma(\cdot)$ is a cost function, C is a constant, and vector w can be written in terms of data points as:

$$w = \sum_{i=1}^{\ell} (\alpha_i - \alpha_i^*) \Phi(x_i). \tag{6.3}$$

By substituting Eq. (6.3) into Eq. (6.1), the generic equation can be rewritten as:

$$f(x) = \sum_{i=1}^{\ell} (\alpha_i - \alpha_i^*)(\Phi(x_i) \cdot \Phi(x)) + b$$

$$= \sum_{i=1}^{\ell} (\alpha_i - \alpha_i^*)k(x_i, x) + b. \tag{6.4}$$

In Eq. (6.4), the dot product can be replaced with function $k(x_i, x)$, which is the kernel function. Kernel is the most important ingredient for SVM. Kernel functions enable dot product to be performed in high-dimensional feature space using low-dimensional space data input without knowing the transformation, Φ. Basically, all kernel functions must satisfy Mercer's condition that corresponds to the inner product of some feature spaces. The radial basis function (RBF) is used in our experiment as the kernel for regression:

$$k(x_i, x) = \exp\left\{-\gamma |x - x_i|^2\right\}. \tag{6.5}$$

Some common kernels are shown in Table 6.1. These are the classical kernels of SVM, which are proposed by Vapnik and his research team members.

The ε-insensitive loss function is the most widely used cost function [70]. The function is in the form:

$$\Gamma(f(x) - y) = \begin{cases} |f(x) - y| - \varepsilon, & for \ |f(x) - y| \geq \varepsilon; \\ 0, & otherwise. \end{cases} \tag{6.6}$$

By solving the following quadratic optimization problem, the regression risk in Eq. (6.2) and the ε-insensitive loss function (6.6) can be minimized:

$$\frac{1}{2}\sum_{i,j=1}^{\ell} (\alpha_i^* - \alpha_i)(\alpha_j^* - \alpha_j)k(x_i, x_j) - \sum_{i=1}^{\ell} \alpha_i^*(y_i - \varepsilon) - \alpha_i(y_i + \varepsilon),$$

subject to

$$\sum_{i=1}^{\ell} \alpha_i - \alpha_i^* = 0, \quad \alpha_i, \alpha_i^* \in [0, C]. \tag{6.7}$$

The Lagrange multipliers, α_i and α_i^*, represent solutions to the above quadratic problem that act as forces for pushing the predictions toward the target value y_i. Only the nonzero values of the Lagrange multipliers in Eq. (6.7) are useful in forecasting the regression line and are called as support vectors. For all points inside the ε-tube,

Table 6.1 Common kernel functions

Kernels	Functions
Linear	$x \cdot y$
Polynomial	$[(x * x_i) + 1]^d$
RBF	$\exp\left\{-\gamma |x - x_i|^2\right\}$

Fig. 6.1 Support vector regression to fit a tube with radius ε to the data and positive slack variables ζ_i measuring the points lying outside of the tube [74]

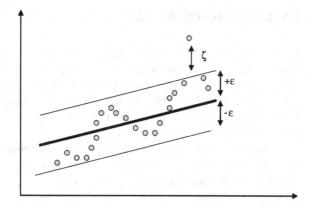

the Lagrange multipliers, equal to zero, do not contribute to the regression function. Only if the requirement $|f(x) - y| \geq \varepsilon$ (as shown in Fig. 6.1) is fulfilled, Lagrange multipliers may assume nonzero values and used as support vectors.

The constant C in Eq. (6.2) determines penalties to estimation errors in the SVR learning process. A large C assigns higher penalties to errors, so that the regression is trained to minimize error with lower generalization, while a small C assigns fewer penalties to errors, which allows the minimization of margin with errors, thus higher generalization ability. If C goes to infinitely large, SVR would not allow the occurrence of any error and would result in a complex model, whereas when C goes to zero, the result would abide a large amount of errors and the model would be less complex.

Now, the value of w in terms of the Lagrange multipliers is deduced. For the variable b, it can be computed by applying Karush–Kuhn–Tucker (KKT) conditions which, in this case, implies that the product of the Lagrange multipliers and constrains has to equal zero:

$$\begin{aligned} \alpha_i(\varepsilon + \zeta_i - y_i + (w, x_i) + b) &= 0, \\ \alpha_i^*\left(\varepsilon + \zeta_i^* + y_i - (w, x_i) - b\right) &= 0, \end{aligned} \tag{6.8}$$

and

$$\begin{aligned} (C - \alpha_i)\zeta_i &= 0, \\ (C - \alpha_i^*)\zeta_i^* &= 0, \end{aligned} \tag{6.9}$$

where ζ_i and ζ_i^* are slack variables used to measure errors outside the ε-tube. Since $\alpha_i, \alpha_i^* = 0$ and $\zeta_i^* = 0$ for $\alpha_i^* \in (0, C)$, b can be computed as follows:

$$\begin{aligned} b &= y_i - (w, x_i) - \varepsilon, \quad \alpha_i \in (0, C); \\ b &= y_i - (w, x_i) + \varepsilon, \quad \alpha_i^* \in (0, C). \end{aligned} \tag{6.10}$$

Putting it all together, SVM and SVR can be used without knowing the transformation.

6.6 Experimental Results

The performance of linear regression, multilayer perceptron, and support vector machine has been tested for the load prediction and demand management. All these algorithms are implemented in Java with default parameter settings, which are available in WEKA [75]. WEKA is a machine learning tool developed at the University of Waikato and has become very popular among the academic community working on learning theory.

In the data modeling environment, modeling is a comparatively easy task rather than predicting an unseen data instance which is called test data. In general, the model prediction accuracy always indicates the strength of the model. Machine learning researchers are considering a range of methods to verify the model strength. Among these, cross-validation [76] is one of the most widely used methods for the final selection of a model. With cross-validation measure, the authors also choose relative absolute error (RAE) [77] to measure the prediction strength for the final model selection. Willmott et al. [78] suggest that mean absolute error (MAE) is a better way for the error measurement in the time series analysis. The mathematical formulation of RAE is presented in the following equation. The range of RAE is between 0 and 100 %. If the value of RAE is 0, it indicates the ideal situation

$$\text{RAE} = \frac{\sum_{i=1}^{N} |p_i - a_i|}{\sum_{i=1}^{N} |\bar{a} - a_i|}, \tag{6.11}$$

where a_i and p_i represent the respective actual and predicted values, respectively, and $\bar{a} = \frac{1}{N}\sum_{i=1}^{N} a_i$ represents the mean of the actual values.

The error measurement in time series analysis is very important because the cost is highly related on the basis of the prediction. By using a simple model, anyone can forecast a string of values for a trend on the basis of the historical time series data. However, right selection of the algorithm is vital for achieve the desired accuracy of the forecast. To examine the seven hours ahead prediction, three well-known algorithms have been selected for the experiment. The data were gathered in an hourly basis. The performances of the algorithms are illustrated in Fig. 6.2.

The forecasting error suggested that SVR was the best choice for the load prediction and demand management tasks. Multilayer perceptron may be considered as a second choice for this task. However, the computational complexity of multilayer perceptron is higher than SVR. Finally, the experimental results suggest that the machine learning techniques are capable of forecasting to implement demand management in the smart grid environment.

Fig. 6.2 Mapping the forecasting model errors of demand management

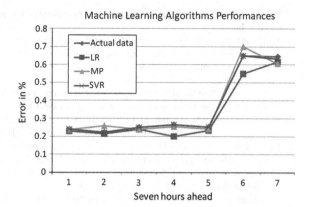

6.7 Conclusions

The chapter presents an overview on demand management state of the art in the smart grid. Some of the popular machine learning algorithms have been tested for the forecasting task to meet the demand management. We consider a real-life data during the performance test of the learning algorithms. In addition, we adopted tenfold cross-validation to make sure for final model selection. From the experimental demonstration, we found that SVR is the best choice to meet our task, which is based on statistical learning theory. SVR is already used for learning to predict future data in many areas. In the forecasting model building process, SVR used a constrained quadratic optimization problem. It implements mapping of inputs onto a high-dimensional space using a set of nonlinear basis functions which is called kernel. This is the specialty of SVR comparing with standard neural networks trained using back-propagation and simple linear regression. However, we need a fully automated SVR algorithm for choosing the kernel function and additional capacity control, development of kernels with invariance.

References

1. IEA (2010) CO_2 emissions from fuel combustion. Highlights. Available via http://www.iea.org/publications/free_new_Desc.asp?PUBS_ID=2143. Cited 5 Jan 2013
2. OECD Factbook (2011–2012) Economic, Environmental and Social Statistics. Available via http://www.oecd-ilibrary.org/sites/factbook-2011-en/06/01/04/index.html?contentType=&itemId=/content/chapter/factbook-2011-49-en&containerItemId=/content/serial/18147364&accessItemIds=&mimeType=text/h. Cited 5 Jan 2013
3. Morgan G, Smuts T (1994) Global warming and climate change. Technical Report, Department of Engineering and Public Policy, Carnegie Mellon University, USA
4. Managing peak electricity demand in South Australia, Available via http://www.sa.gov.au/subject/Water,+energy+and+environment/Energy/Energy+supply,+providers+and+bills/

Electricity+and+gas+supply/Managing+peak+electricity+demand+in+South+Australia.
Cited 6 Jan 2013

5. Molderink A, Bakker V, Bosman M, Hurink J, Smith G (2010) A three-step methodology to improve domestic energy efficiency. In: Proceedings of IEEE PES conference on innovative smart grid technologies, pp 1–8

6. Mohsenian-Rad A-H, Wong VW, Jateskevich J, Schober R, Leon-Garcia A (2010) Autonomous demand-side management based on game-theoretic energy consumption scheduling for the future grid. IEEE Trans Smart Grid 1(3):320

7. Medina J, Muller N, Roytelman I (2010) Demand response and distribution grid operations: opportunities and challenges. IEEE Trans Smart Grid 1(2):193–198

8. Heinemann G, Nordman D, Plant E (1996) The relationship between summer weather and summer loads: a regression analysis. IEEE Trans Power Apparatus Syst (PAS) 85:1144–1154

9. Hahn J, Meyer-Nieberg S, Pickle S (2009) Electric load forecasting methods: tools for decision making. Eur J Oper Res 199:902–907

10. Erol-Kantarci M, Mouftah HT (2011) Demand management and wireless sensor networks in the smart grid, energy management systems. In: Giridhar Kini P (ed) ISBN: 978-953-307-579-2. Available at: http://www.intechopen.com/books/energy-management-systems/demand-management-and-wireless-sensor-networks-in-the-smart-grid

11. Gross G, Galiana FD (1987) Short-term load forecasting. Proc IEEE 75(12):1558–1573

12. Feinberg EA, Genethliou D (2006) Load forecasting, applied mathematics for restructured electric power systems. Springer, Berlin

13. Mohsenian-Rad AH et al (2010) Autonomous demand-side management based on game-theoretic energy consumption scheduling for the future smart grid. IEEE Trans Smart Grid 1(3):320–331

14. Charytoniuk W et al (1998) Nonparametric regression based short-term load forecasting. IEEE Trans Power Syst 13:725–730

15. Box GEP, Jenkins GM (1970) Time-series analysis: forecasting and control. Holden-Day, San Francisco

16. Gupta PC (1985) Adaptive short-term forecasting of hourly loads using weather information. In: Bunn DW, Farmer ED (eds) Comparative models for electrical load forecasting. Wiley, New York, pp 43–56

17. Chen T (1997) Long-term peak electricity load forecasting in Taiwan: a cointegration analysis. Pac Asian J Energy 7(1):63–73

18. Soares LJ, Souza LR (2006) Forecasting electricity demand using generalized long memory. Int J Forecast 1(22):17–28

19. Soares LJ, Medeiros MC (2005) Modelling and forecasting short-term electric load demand: a two-step methodology. Working Paper 495, Department of Economics, Pontifical Catholic University of Rio de Janeiro

20. Soares LJ, Medeiros MC (2008) Modeling and forecasting short-term electricity load: a comparison of methods with an application to Brazilian data. Int J Forecast 24:630–644

21. Amaral LF et al (2008) A smooth transition periodic autoregressive (STPAR) model for short-term load forecasting. Int J Forecast 24:603–615

22. Hippert HS et al (2001) Neural networks for short-term load forecasting: a review and evaluation. IEEE Trans Power Syst 16(1):44–55

23. Hippert HS, Pedreira CE (2004) Estimating temperature profiles for short term load forecasting: neural networks compared to linear models. IEE Proc Gener Transm Distrib 151(4):543–547

24. Heinemann GT et al (1996) The relationship between summer weather and summer loads—a regression analysis. IEEE Trans Power Apparatus Syst 85(11):1144–1154

25. Krogh B et al (1982) Design and implementation of an on-line load forecasting algorithm. IEEE Trans Power Apparatus Syst 101(9):3284–3289

26. Papalexopoulos AD, Hesterberg TC (1990) A regression-based approach to short-term system load forecasting. IEEE Trans Power Syst 5(4):1535–1547

27. Baran ME et al (2005) Load estimation for load monitoring at distribution substations. IEEE Trans Power Syst 20(1):164–170
28. Hong T, et al (2010) Modeling and forecasting hourly electric load by multiple linear regression with interactions. In: Proceedings of 2010 IEEE PES general meeting
29. Urso PD (2003) Linear regression analysis for fuzzy/crisp input and fuzzy/crisp output data. Comput Stat Data Anal 42:47–72
30. Kaur M, Kumar A (2012) Fuzzy optimal solution for unbalanced fully fuzzy minimal cost flow problems. Int J Fuzzy Syst 14(1):1–10
31. Tanaka H, Watada J (1988) Possibilistic linear systems and their application to the linear regression model. Fuzzy Sets Syst 27(2):275–289
32. Peters G (1994) Fuzzy linear regression with fuzzy intervals. Fuzzy Sets Syst 63:45–55
33. Redden DT, Woodall WH (1994) Properties of certain fuzzy linear regression methods. Fuzzy Sets Syst 64:361–375
34. Donoso S et al (2006) Quadratic programming models for fuzzy regression. In: Proceedings of international conference on mathematical and statistical modeling in honor of Enrique Castillo
35. Bárdossy A (1990) Note on fuzzy regression. Fuzzy Sets Syst 65:65–75
36. Diamond P (1988) Fuzzy least squares. Inf Sci 46:141–157
37. McMenamin JS (1997) A primer on neural networks for forecasting. J Bus Forecast Methods Syst 16(3):17–22
38. Alfares HK, Nazeeruddin M (2002) Electric load forecasting: literature survey and classification of methods. Int J Syst Sci 33(1):23–34
39. Khotanzad A et al (2000) Combination of artificial neural network forecasters for prediction of natural gas consumption. IEEE Trans Neural Networks 11:464–473
40. Weron R, Misiorek A (2004) Modeling and forecasting electricity loads: a comparison. In: Proceedings of the European electricity market EEM-04, pp 135–142
41. Kavaklioglu KH et al (2009) Modeling and prediction of Turkey's electricity consumption using artificial neural networks. Energy Convers Manage 50:2719–2727
42. Babu PR et al (2007) Neural network and DSM techniques applied to a industrial consumer: a case study. In: Proceedings of 5th IEEE international conference on compatibility in power electronics, pp 1–4
43. Wong S et al (2010) Artificial neural networks for energy analysis of office buildings with daylighting. Appl Energy 87(2):551–557
44. Fadare D (2009) Modelling of solar energy potential in Nigeria using an artificial neural network model. Appl Energy 86:1410–1422
45. Haykin S (2009) Neuronal networks and learning machines, 3rd edn. Pearson International, New Jersey
46. Ravi PB et al (2008) Application of ANN and DSM techniques for peak load management: a case study. In: Proceedings of the IEEE region 8 international conference on computational technologies in electrical and electronics engineering
47. Atwa YM et al (2007) DSM approach for water heater utilizing Elman neural network. In: Proceedings of the IEEE electrical power conference
48. Musilek P et al (2006) Recurrent neural network based gating for natural gas load prediction system. In: Proceedings of the international conference on neural networks (IJCNN'06), pp 3736–3741
49. Bakirtzis A et al (1995) Short term load forecasting using fuzzy neural networks. IEEE Trans Power Syst 10:1518–1524
50. Li Z et al (2007) Neural network prediction of energy demand and supply in China. In: Proceedings of the institution of civil engineers—energy, vol 160, pp 145–149
51. Lee KY et al (1991) A Study on neural networks for short-term load forecasting. In: Proceedings of the 1st forum on application of neural networks to power systems, pp 26–30
52. Park DC et al (1991) Electric load forecasting using an artificial neural network. IEEE Trans Power Syst 6(2):442–449

53. Cristianini N, Shawe-Taylor J (2000) An introduction to support vector machines and other kernel-based learning methods, 1st edn. Cambridge University Press, UK
54. Khan MS, Coulibaly P (2006) Application of support vector machine in lake water level prediction. J Hydrol Eng 11(3):199–205
55. Pires M, Marwala T (2004) Option pricing using neural networks and support vector machines. In: Proceedings of the IEEE international conference on systems, man and cybernetics, The Hague, Nederland, pp 1279–1285
56. Jayadeva RK, Chandra S (2009) Regularized least squares fuzzy support vector regression for financial time series forecasting. Expert Syst Appl 36(1):132–138
57. Zhang J et al (2008) A pattern recognition technique for structural identification using observed vibration signals: nonlinear case studies. Eng Struct 30:1417–1423
58. Mukherjee S et al (1997) Nonlinear prediction of chaotic time series using support vector machines. In: Proceedings of the IEEE NNSP'97, pp 24–26
59. Mohandes MA et al (2004) Support vector machines for wind speed prediction. Renew Energy 29:939–947
60. Tay FEH, Cao L (2001) Application of support vector machines in financial time series forecasting. Omega: Int J Manage Sci 29:309–317
61. Cao L, Gu Q (2002) Dynamic support vector machines for non-stationary time series forecasting. Intell Data Anal 6:67–83
62. Tay FEH, Cao L (2002) Modified support vector machines in financial time series forecasting. Neurocomputing 48:847–861
63. Wang W et al (2003) Three improved neural network models for air quality forecasting. Eng Comput 20:192–210
64. Zhou Q et al (2007) Sequential minimal optimization algorithm applied in short-term load forecasting. In: Proceedings of the 6th international conference on machine learning and cybernetics, pp 2479–2483
65. Li W et al (2009) Study on long-term load forecasting of MIXSVM based on principal component analysis. In: Proceedings of 2009 international conference on future biomedical information engineering, pp 439–441
66. Li W et al (2010) Study on long-term load forecasting of MIX-SVM based on rough set theory. Power Syst Protect Control 38(13):31–34
67. Pai P-F et al (2006) A hybrid support vector machine regression for exchange rate prediction. Inf Manage Sci 17(2):19–32
68. Box GEP et al (1994) Time series analysis, forecasting and control, 3rd edn. Prentice Hall, NJ
69. Stock JH, Watson MW (2003) Introduction to econometrics. Pearson India, New Delhi
70. Vapnik V (1995) The nature of statistical learning theory. Springer, NY
71. Burges C, (1998) A tutorial on support vector machines for pattern recognition. In Trybula WJ (ed) Data mining and knowledge discovery, vol 2. Kluwer Academic Publishers, Boston
72. Vapnik V et al (1997) Support vector method for function approximation, regression estimation, and signal processing. In Mozer M, Jordan M, Petsche T (eds) Advances in neural information processing systems, vol 9. MIT Press, Cambridge, pp 281–287
73. Chatfield C (1996) The analysis of time series—an introduction. Chapman and Hall, London
74. Wu C-H et al (2004) Travel-time prediction with support vector regression. IEEE Trans Intell Transp Syst 5:276–281
75. Witten IH et al (2011) Data mining: practical machine learning tools and techniques. Morgan Kaufmann, U.S
76. Contextuall (2012) What is 10-fold cross validation? Available via https://contextuall.com/what-is-10-fold-cross-validation/. Cited 12 Jan 2013
77. Cohen J (1960) A coefficient of agreement for nominal scales. Educ Psychol Measur 20:37–46
78. Willmott CJ, Matsuura K (2005) Advantages of the mean absolute error (MAE) over the root mean square error (RMSE) in assessing average model performance. Clim Res 30:79–82

Chapter 7
Database Systems for the Smart Grid

Zeyar Aung

Abstract In this chapter, two aspects of database systems, namely database management and data mining, for the smart grid are covered. The uses of database management and data mining for the electrical power grid comprising of the interrelated subsystems of power generation, transmission, distribution, and utilization are discussed.

7.1 Introduction

Since the smart grid rely on modern information and communication technology (ICT) infrastructure, database systems, which are one of the vital components of ICT, are indispensable in the smart grid. Database systems allow the data in the smart grid to be stored in a systematic manner and enable them to be retrieved, processed, and analyzed either immediately (i.e., online data processing/analysis) or later (i.e., historical data processing/analysis).

Because of the involvement in database systems, the smart grid is no longer a business dominated by utility companies and electricity hardware companies alone. Several big software companies in data-centric business such as Teradata [1], Oracle [2], SAS [3], SAP [4], IBM [5], Microsoft [6], and Google [7] are active players in the smart grid arena now.

There are two main aspects of a database system, namely database management (data storage, transaction processing, and querying) and data mining (analysis of data to gain certain knowledge or facilitate certain decision making). These two aspects are naturally interrelated and are like the two sides of a coin. Both are essential for the business process of the smart grid's operations.

Z. Aung (✉)
Computing and Information Science Program, Masdar Institute of Science and Technology,
Block 3, Masdar City, Abu Dhabi 54224, United Arab Emirates
e-mail: zaung@masdar.ac.ae

A. B. M. S. Ali (ed.), *Smart Grids*, Green Energy and Technology,
DOI: 10.1007/978-1-4471-5210-1_7, © Springer-Verlag London 2013

In this chapter, we will cover the applications of database management and data mining in the smart grid for power generation, transmission, distribution, and utilization (consumption). Again, these four application areas are interrelated and somewhat overlapping, especially because of the interconnected nature of the smart grid.

The development of smart grid is an evolutionary process. During the smart grid's introduction phase, the two generations of technologies will coexist [8]. For ICT components (both software and hardware), a majority of legacy systems are first to be integrated into the smart grid and later phased out and replaced by the newer technologies. However, for power system components, the introduction of smart gird will not even drastically change the basic mechanisms of the power system's mechanical and electrical equipment (except that they will now be more intelligent and responsive because of incorporation of ICT). For example, a gas turbine will still operate just in the same way to convert natural gas into electrical power as it did in the old non-smart grid — albeit it may now use less amount of gas because of a more intelligent control system. So, a database recording the operations of such a gas turbine will be more or less the same in both the traditional grid and the smart grid.

For the aforementioned reasons, we believe that both the earlier systems for systematic power grid data management/data mining even before the word smart grid was coined and the newer systems which were explicitly proposed for the smart grid are worth covering. As such, in this chapter, we will include the literature on power grid database systems both before and after the concept of the smart grid was conceived.

In the following two sections, database management and data mining for power grids will be, respectively, covered.

7.2 Power Grid Database Management

In this section, we will cover the database management technologies in general and then the applications of database management for a power grid in its four subsystems: generation, transmission, distribution, and utilization.

7.2.1 Database Management Technologies

In modern days, management of data in an ICT system is centered around a proper database management system (DBMS) or sometimes simply a file system (FS). In both cases, the basic operations of data management are as follows:

- Schema creation: defining format of data and relationships among data.
- Data insertion: populating the database/files with data.

- Data maintenance: updating or deleting existing data.
- Querying and reporting: retrieval of stored data as per users' business requirements.
- Performance optimization: making the retrieval process faster by using indexes, etc.
- User account management: defining which user has a right to do which operations on which data.
- Backup and recovery: preventing accidental loss of data.

For DBMS, relational database (composing of tables which are mathematically termed "relations") is the most common standard. Some commonly used relational DBMSs are Oracle (proprietary), Microsoft SQL Server (proprietary), IBM DB2 and Informix (proprietary), SAP Sybase (proprietary), MySQL (open source), and PostgreSQL (open source). Structured query language (SQL) is a common interface to retrieve data from relational DBMS.

Recently, post-relational database systems called NoSQL (Not only SQL) [9] become more and more common. NoSQL database systems include document-oriented databases (e.g., MongoDB), XML databases (e.g., BaseX), graph databases (e.g., InfiniteGraph), key-value stores (e.g., Apache Cassandra), multi-value databases (e.g., OpenQM), object-oriented databases (e.g., db4o), RDF (resource description framework) databases (e.g., Meronymy SPARQL), tabular databases (e.g., BigTable), tuple databases (e.g., Jini), and column-oriented databases (e.g., c-store). NoSQL database systems use conventional programming languages like C++, C#, Java, and Erlang, or XQuery in the case of XML databases in order to interface and retrieve data from the databases.

In addition to NoSQL databases, parallel and distributed file systems such as Apache Hadoop [10] and Google MapReduce [11] become increasingly popular. Since the smart grid by its own nature is distributed and the resources (like smart meters, meter data concentrators, substation transformers) in it are geographically scattered, distributed file systems can potentially be very useful for the smart grid.

Generally, databases are stored on centralized or distributed magmatic hard disk drives. However, new paradigms of databases stored on main memory (such as voltDB) and solid-state drives (such as [12]) are emerging because of the increased availability of high-capacity main memory and solid-state equipment at low costs.

Another increasing popular approach nowadays is to store databases in the cloud. Cloud computing and cloud database [13] are also the emerging trends that are much relevant to the smart grid. A cloud database can be in the form of either a virtual machine instance which can be purchased for a limited time or a database as a service in which the service provider installs and maintains the database, and application owners pay according to their usage. Amazon's DynamoDB and SimpleDB are examples of database as a service.

"Big data" (meaning several tera- to petabytes of data) is one of the current hot topics. Big data is a crucial issue for the smart grid since an enormous volume of data is expected to be generated from its large number of connected devices and sensors at every short time interval. IBM Netezza is one of the examples of DBMS that can handle big data. The parallel/distributed data management techniques of Hadoop and MapReduce are also highly relevant to deal with big data because usually the big data is not centralized but distributed among several computing resources.

Finally, data integration is an important issue for complex systems with multiple components like the smart grid. Data from different sources, probably by different vendors, having different formats and semantics are to be systematically integrated to form a single uniform data source, which can be either virtual or physical. Such an integrated data source can facilitate an integrated information system that streamlines various business processes in a utility company. Most common data integration techniques are data warehousing, XML, and ontology-based techniques.

A high-level diagram illustrating the interrelationships among the various modern database management technologies and their applications in the different areas of power grid data management is shown in Fig. 7.1.

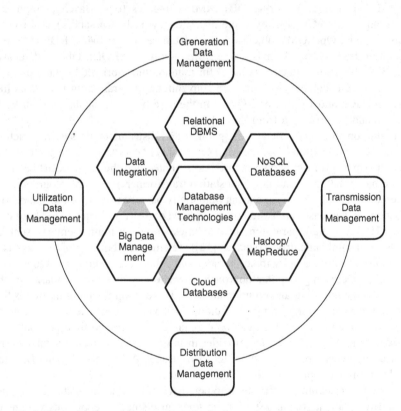

Fig. 7.1 Modern database management technologies and their applications in power grid database management.

7.2.2 Generation Data Management

Power plants generate electrical power from traditional sources such as natural gas, petroleum, coal, nuclear or hydropower as well as modern renewable sources such as wind or solar power. Database systems for power plants have different structures and contents depending on the type of the energy source.

Li et al. [14] describe a database system for a coal-based power plant which records and processes the data specific to coal-operated stream turbines (such as main steam pressure, feed water temperature, reheater spray, flue gas temperature, excess air coefficient, and condenser vacuum).

Huang et al. [15] discuss data management systems for a hydropower plant, particularly the automatic generator tripping and load shedding system installed at the Churchill Falls hydropower plant in Labrador, Canada by Hydro-Quebec.

Swartz et al. [16] propose a wireless sensor network-based data collection and management system for wind farms to provide information about the dynamic behavior of the wind turbines and their response to loading.

7.2.3 Transmission and Distribution Data Management

After the power has been generated, it is transformed into high-voltage electricity using step-up transformers and is transmitted along the transmission lines to multiple substations. At a substation, the electricity's voltage is transformed again to a level suitable for consumption by using a step-down transformer. Then, the electricity is distributed to the consumers for utilization.

Early examples of database systems for power transmission/distribution systems in the literature are [17] and [18].

Generally, distributed control system (DCS) and supervisory control and data acquisition (SCADA) are employed to operate various equipment used in power transmission and distribution. DCS and SCADA are usually proprietary systems from big industrial players in the power industry such as GE [19] and Siemens [20]. Being proprietary systems, they are closed and sometimes can be legacy systems. In some cases, the data format they provide can be non-standard (especially for old legacy systems). Thus, acquiring data from all these systems to build a common database system can be sometimes difficult. In the worst cases, manual data entry can be required [21].

It is not uncommon to have systems from multiple vendors in a single power facility. In order to provide a standardized interface and allow easy exchange of data among different prices of software by different vendors, common information model (CIM) [22],[23], generic substation events [24], and substation configuration language (SCL) [25] have been proposed.

Depending on the nature of application, the data generated by various pieces of power system equipment have to be stored in different formats [26]. They are as follows:

- Raw waveforms (voltage and currents) sampled at relatively high sampling frequencies.
- Pre-processed waveforms (e.g., RMS) typically sampled at low sampling frequencies.
- Status variables (e.g., if a relay is opened or closed) typically sampled at low sampling frequencies.

A number of white papers and research articles on the database systems for power transmission/distribution systems exist in the literature. Some examples, which are by no means complete, are as follows.

Simpson [21] describes a power system database recording transformer name-plate data, single line diagrams, measured data, protective device coordination, harmonic analysis, transistent calculation, load flow calculation, and short-circuit calculation. Martinez et al. [27] give detailed descriptions about comprehensive archiving and management of power system data for real-time performance monitoring using CERTS (Consortium for Electric Reliability Technology Solutions) architecture. Qiu et al. [28] propose a system of real-time and historical (archived) databases to allow operations, controls, and analysis of power transmission and distribution. An example of a practical database schema to be used for in transmission utility enterprise-wide framework using ArcGIS, ArcSDE, Microsoft SQL Server, and .NET is given in [29]. In [30] and [31], the issues of data integration in power systems are discussed. In [32], Zheng et al. propose a cloud computing and cloud database framework for substations of the smart gird. Rusitschka et al. [33] discuss the use of cloud data management for outage management [34] and virtual power plant [35].

A comprehensive list of monitoring subsystems whose measurement data are to be collected and stored in the database for a modern power transmission/distribution system of the smart grid is provided by Kaplan et al. [36]. These collected data allow advanced tools to analyze system conditions, perform real-time contingency analysis, and initiate a necessary course of action as needed. These monitoring subsystems as described in [36] are

- **Wide-area monitoring system:** GPS (global positioning system)-based phasor monitoring unit (PMU) that measures the instantaneous magnitude of voltage or current at a selected grid location. This provides a global and dynamic view of the power system.
- **Dynamic line rating technology:** it measures the ampacity of a line in real time.
- **Conductor/ compression connector sensor:** it measures conductor temperature to allow accurate dynamic rating of overhead lines and line sag, thus determining line rating.

- **Insulation confirmation leakage sensor:** it continuously monitors leakage current and extracts key parameters. This is critical to determining when an insulator flashover is imminent due to contamination.
- **Backscatter radio:** it provides improved data and warning of transmission and distribution component failure.
- **Electronic instrument transformer**: it replaces precise electromagnetic devices (such as current transformers and potential transformers) that convert high voltages and currents to manageable, measurable levels.
- **Other monitoring systems:**

  - Fiber-optic, temperature monitoring system.
  - Circuit breaker real-time monitoring system.
  - Cable monitor.
  - Battery monitor.
  - Sophisticated monitoring tool which combines several different temperature and current measurements.

7.2.4 Utilization Data Management

The distributed electricity is utilized (consumed) at the consumers' end. Consumers may be of several types: residential (e.g., individual houses and apartment buildings), commercial (e.g., banks), industrial (e.g., factories), transportation (e.g., subways), emergency services (e.g., hospitals), and governmental services (e.g., police), etc. Obviously, power utilization is most visible aspect of a power grid for the public.

In the old traditional grid, a traditional meter on customer's premises is read by a meter reading staff at a regular interval (e.g., once a month), and the meter readings (utilization data) are manually entered into the database system in the utility company. These utilization data are quite passive and are mainly used for the purpose of billing. It has no or little use in real-time monitoring and control of the power system in operation because of a very long time lag (e.g., up to one month) between actual power utilization and data gathering.

However, in the era of the smart grid, smart meters are installed in consumers' premises. Among its many functionalities, the main function of a smart meter is to record and transmit the utilization data to the utility company at relatively short time intervals (e.g., every 5, 10, or 15 minutes). The utilization data can be either fine-grained (separate data for individual appliances or groups of appliances in the same electrical circuits) or coarse-grained (aggregated data for the whole premises). A smart meter may be equipped with a small local storage (e.g., SD card) to store some intermediate utilization data.

The data collection is hierarchical in nature. The power utilization data from a number of smart meters are first transmitted to a data concentrator, and a number of data concentrators relay the data to the central server at the utility company where the data are stored in the utilization database covering a large number of consumers.

The above process of data collection is called automatic meter reading (AMR) [37]. It is a one-way communication process in which the data are transmitted from the smart meter end to the server end through the data concentrator. Later, AMR is improved into a more sophisticated system named advanced metering infrastructure (AMI) [38],[39]. AMI allows two-way communication between the smart meter and the server end. The server can send messages regarding real-time pricing, control commands to switch on/off certain appliances, etc. to the smart meter.

In a smart home environment, where modern technologies such as smart appliances, intelligent heating, ventilation, air conditioning (HVAC), rooftop solar generation, and electric/hybrid vehicles coexist, a smart meter alone will not be able to handle all the data regarding the operations and interactions among those equipment. In addition to the smart meter, there requires a local PC/server to host an integrated information management platform. Its purpose is to store, process, and manage the data from all those smart installations and to communicate with the utility to exchange the relevant information regarding them. Lui et al. [40] describe in detail such a platform namely Whirlpool integrated services environment (WISE), which is a proprietary system.

Since every customer connected to the smart grid is expected to generate a large volume of data from his/her smart meter as well as from the other multiple smart equipments, there is a pressing need for the smart grid to handle the big data (as also discussed above in Sect. 7.2.1). In [41], the application of IBM's big data technologies for smart meters is discussed.

Kaplan et al. [36] provide the following detailed list of customer-focused applications (for each of which the relevant utilization data are needed to be recorded and processed).

- **Consumer gateway:**
  - Bidirectional communications between service organizations and equipment on customer premises.
  - Advanced meter reading.
  - Time-of-use and real-time pricing (RTP).
  - Load control.
  - Metering information and energy analysis via website.
  - Outage detection and notification.
  - Metering aggregation for multiple sites and facilities.
  - Integration of customer-owned generation.
  - Remote power quality monitoring and services.
  - Remote equipment performance diagnosis.
  - Theft control.
  - Building energy management systems.
  - Automatic load controls integrated with RTP.

- Monitoring of electrical consumption of total load and, in some cases, various load components.
- Functions embodied in meters, cable modems, set-top boxes, thermostats, etc.

- **Residential consumer network:** subset of consumer gateway concept.

  - Reads the meter, connects controllable loads, and communicates with service providers.
  - End users and suppliers monitor and control the use and cost of various resources (e.g., electricity, gas, water, temperature, air quality, secure access, and remote diagnostics).
  - Consumers monitor energy use and determine control strategies in response to price signals.

- **Advanced meter:**

  - Employs digital technology to measure and record electrical parameters (e.g., watts, volts, and kilowatt hours).
  - Communication ports link to central control and distributed loads.
  - Provides consumption data to both consumer and supplier.
  - Switches loads on and off in some cases.

At the utility side, billing is the most important application for the utilization data. Arenas-Martinez et al. [42] developed a smart grid simulation platform to study the pros and cons of different database architectures for massive customer billing. These architectures are single relational database, distributed relational database, key-value distributed database storage, and hybrid storage (DBMS and FS).

Another utility-side application relying on the utilization data is real-time pricing to facilitate demand response by having the consumers reduce their demand at critical times or in response to market prices [43].

7.3 Power Grid Data Mining

In this section, we will cover the data mining technologies in general and then applications of data mining for a power grid in its four subsystems: generation, transmission, distribution, and utilization.

7.3.1 Data Mining Technologies

The purpose of data mining is to uncover the knowledge or interesting patterns of data that lie within a large database and use them for decision support at various levels (strategic, tactical, or operational). Data mining is also known by other names such as data analytics, knowledge discovery, and statistical data analysis.

Data mining is closely related to database management, machine learning, artificial intelligence, and statistics.

The most common data mining tasks are

- **Frequent pattern mining:** to discover some subpatterns or motifs that occur frequently in a dataset. (Note: a dataset means a collection of data organized in rows and columns. It can be a table in relational DBMS or just a comma-separated values (CSV) file in FS. A row represents an instance, and a column represents an attribute.) Some well-known frequent pattern mining algorithms include *a priori*, *FP-tree*, and *Eclat*.
- **Association rule mining:** to uncover which causes usually lead to which effects in a dataset. The association rules can generally be derived from the frequent patterns described above.
- **Classification:** to classify instances in a dataset into pre-defined groups (called class labels). Classification is a supervised learning process in which we first have to train the classifier with instances whose class labels are known. Then, we use this training classifier to predict the class labels of the new instances whose labels are not known yet. Some popular classification algorithms are *decision tree, naive Bayes, artificial neural networks, hidden Markov model, support vector machine*, and *k-nearest neighbors*.
- **Clustering:** to organize similar instances in a dataset into groups which are not pre-defined. Clustering is an unsupervised learning process in which we do not know the class labels of all the instances in the dataset in advance. The number of groups (clusters) may or may not be pre-defined, depending on the clustering algorithm. Some widely used clustering algorithms are *k-means, fuzzy c-means, expectation maximization, DBSCAN, BIRCH*, and *hierarchical clustering*.
- **Regression:** to predict the value of the target attribute (called dependent variable) of an instance based on the values of other attributes (independent variables). Regression is also a type of supervised learning which works in the similar way as classification. Their main difference is that while the outputs of classification are class labels (discrete values), those of regression are real numbers (continuous values). Some common regression algorithms are *Gauss–Newton algorithm, logistic regression, neural network regression, support vector regression*, and *autoregressive integrated moving average (ARIMA)*.
- **Outlier detection:** to identify anomalous instances, which might be interesting, or indicate errors that require further investigation. It can be supervised, unsupervised, or semi-supervised learning. Some popular methods are *local outlier factor, single-class support vector machine, replicator neural networks*, and *cluster analysis*.

Data can rarely be mined in their raw forms as originally stored in the DBMS or FS. We usually need to perform one or more of the following data processing tasks [44] before performing a data mining task.

- **Data cleaning:** to fill in missing values, smooth noisy data, identify or remove outliers, and resolve inconsistencies.

- **Data integration:** to integrate multiple databases, data cubes, or files.
- **Data reduction:** to obtain reduced representation in volume but produces the same or similar analytical results. It may be in the form of dimensionality reduction, numerosity reduction, or data compression. Data reduction is usually done for the sake of efficiency and/or better generalization.
- **Data transformation and discretization:** to normalize data, aggregate data, and generate concept hierarchy.

After the data mining task has been performed, the result can be optionally presented in a visual format in order to better facilitate decision making by the user.

Some popular data mining software are SAS Enterprise Miner (proprietary), IBM SPSS Modeler (proprietary), Oracle Data Mining (proprietary), Microsoft Analysis Services (proprietary), Weka (open source), RapidMiner (open source), and ELKI (open source).

In addition to the traditional data mining paradigm on static and centralized data, the new paradigms of distributed data mining [45], data stream mining [46], and time series data mining [47] are much relevant to the smart grid because of its very nature of distributiveness and having to deal with numerous data streams and time series data from various data sources: smart meters, sensors, and power system machinery.

Privacy is one of the top concerns in the smart grid's deployment, especially from consumer's perspective [48]. Thus, privacy-preserving data mining techniques [49] are much relevant for mining the data in the smart grid. An example of a proposed framework for privacy-preserving data integration and subsequent analysis for the smart grid is [50].

A high-level diagram depicting the interrelationships among the various data mining technologies and their applications in the different subsystems of power grids is shown in Fig. 7.2.

7.3.2 Data Mining for Generation

In a similar manner as discussed above in Sect. 7.2.2, the data mining applications for power generation can be quite diverse because of the different natures of power sources. Li et al. [14] propose a fault diagnosis system for a coal-based power plant using association rule mining. In [51], the operational performance and the efficiency characteristics for photovoltaic power generation are analyzed against various environmental conditions using statistical analysis.

For fossil fuel-based power plants where the amount of power produced can be fully controlled, the amount of generation (supply) is much dependent on the amount of electricity load (demand). So, forecasting the future load enables them to plan for the required fuel accordingly, and consequently, accurate forecasting can save utility companies millions of dollars a year [52]. Also, for renewable energy generations, load forecasting can help the utilities to plan ahead to shave

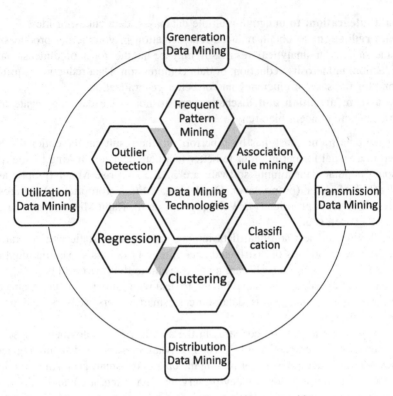

Fig. 7.2 Data mining technologies and their applications in power grids.

the peak load by means of demand response mechanisms [43] so that the demand will not exceed the available power output from the renewable source.

Load forecasting can be for very-short term (24 hours ahead of the present time), short term (~ 2 weeks), medium term (~ 3 years), and long term (~ 30 years) [54]. Some examples of load forecasting methods in the literature are Deng and Jirutitijaroen [53] using the time series models of multiplicative decomposition and seasonal ARIMA, Hong [54] using multiple linear regression, Zhang et al. [55] using artificial neural network, and Aung et al. [56] using least square support vector regression. Taylor [57] provides a good survey and evaluation of several existing load forecasting methods.

7.3.3 Data Mining for Transmission and Distribution

The prospects and challenges of data mining for the smart grid, particularly in the areas of transmission, distribution, and utilization, are highlighted in [58]. Similarly, Ramchurn et al. [59] discuss the uses of artificial intelligence and data mining solutions to provide "smartness" to the smart grid.

There exists a number of papers in the literature regarding the application of data mining for power transmission and distribution systems. Some examples, which are by no means exhaustive, are as follows.

Dissolved gas analysis (DGA) [60] is the study of dissolved gases in transformer oil (insulating oil which is stable at high temperatures and possesses excellent electrical insulating properties). The information about the gases being generated by a particular transformer unit can be very useful in fault detection and maintenance. Sharma et al. [61] provide a survey on artificial intelligence and data mining techniques for DGA.

Power system state estimation provides an estimate for all metered and unmetered quantities throughout the whole power system. It is useful in ensuring the stability of the grid and preventing blackouts. Chen et al. [62] describe computation of power system state estimation using weighted least square method on a high-performance computing platform. Zhong et al. [63] try to solve a more specific problem of state assessment for transformer equipment using association rule mining and fuzzy logic.

Islanding detection is also important for the stability of a grid in which multiple small distributed renewable energy generation sources are integrated into the main grid. Islanding occurs when part of the network becomes disconnected from the grid and is powered by one or more distributed generations only. Such an event can potentially lead to problems in the grid. Samantaray et al. [64] proposed an islanding detection system using a rule-based approach that employs fuzzy membership functions. In [65], naive Bayes classifier is used to solve the problem of islanding detection.

Again, fault identification and fault cause identification are obviously important problems for power systems. Calderaro et al. [66] uses Petri Nets to solve the fault identification problem. Xu et al. [67] try to identify fault causes in a power distribution system using a fuzzy classification algorithm.

Contingency analytics is to understand the impact of potential component failures and assess the power system's capability to tolerate them. Adolf et al. [68] develop a filtering technique based on multi-criteria optimization to address it.

Power quality is another important issue in the power system, especially in the smart grid era. Common problems that can disturb the quality of power are sags (undervoltages), harmonics, spikes, and imbalances [36]. He et al. [69] propose a self-organizing learning array system for power quality classification based on wavelet transform. Hongke and Linhai [70] describe a practical data analysis platform for power quality using Microsoft SQL Server and OLAP (online analytical processing).

The reliability of the power distribution network is an important issue, especially for the old networks that were first setup nearly a century ago. Gross et al. [71] develop a support vector machine–based model to rate the feeder lines in New York City for their reliability and identify the ones that need maintenance or replacement.

Morais et al. [26] present a good survey of 13 research articles on data mining for power systems for various purposes such as fault classification and location, detection and diagnosis of transient faults, power quality detection for power system disturbances. Similarly, Mori [72] provides a list of 42 research papers on various applications of data mining for power systems.

Apart from the physical power system, the logical energy market draws much attention recently, especially after its deregulation. Price forecasting is an indispensable tool for both the energy wholesaler and the retailer in such a market. Arenas-Martinez et al. [73] present a price forecasting model using local sequence patterns, while Neupane et al. [74] tackle price forecasting by means of artificial neural networks.

7.3.4 Data Mining for Utilization

At the power utilization (demand) side, load forecasting for large commercial and residential buildings plays a crucial role. Building load forecasting is an integral part of a building management system. It enables the building operator to plan ahead, shave loads if required, and carry out fault identification and diagnosis in the building's electrical system if necessary. Fernandez et al. [75] present a study on building load forecasting using autoregressive model, polynomial model, neural network, and support vector machine. Edwards et al. [76] compare the performance of seven machine learning/data mining methods for load forecasting in buildings.

Customer profiling is also related to the demand-side load forecasting task mentioned above. It is useful both for customer behavior prediction for appliance scheduling automation and for dynamic pricing of electricity to suit individual customers' usage patterns. Proposed research works for customer profiling using data mining techniques include [77, 78], and [79].

Finally, security is one of the major concerns for the smart grid's deployment at the customer side [80]. To partially address this problem, Faisal et al. [81] present an intrusion detection system for advance metering infrastructure (AMI) using data stream mining methods. Fatemieh et al. [82] apply classification techniques to improve the attack resilience of TV spectrum data fusion for AMI communications.

7.4 Conclusion

Database systems are one of the keystones of the ICT infrastructure that provides smartness to the smart gird. In this chapter, we have discussed both the conventional and the state-of-the-art database system technologies regarding database management and data mining and their applications to the smart grid. We hope our chapter to be useful as a reference material for both the researchers and the practitioners of the smart grid.

Acknowledgments The author thank the Government and Abu Dhabi, United Arab Emirates, for sponsoring this research through its funding of Masdar Institute–Massachusetts Institute of Technology (MIT) collaborative research project titled "Data Mining for Smart Grids".

References

1. Teradata Corporation. http://www.teradata.com
2. Oracle Corporation. http://www.oracle.com
3. SAS Institute, Inc. http://www.sas.com
4. SAP AG http://www.sap.com
5. IBM corporation. http://www.ibm.com
6. Microsoft Corporation. http://www.microsoft.com
7. Google Inc. http://www.google.com
8. Farhangi H (2010) The path of the smart grid. IEEE Power Energy Mag 8:18–28
9. Wikipedia: NoSQL (2013). http://en.wikipedia.org/wiki/NoSQL
10. Apache Hadoop. http://hadoop.apache.org
11. Wikipedia: MapReduce (2013). http://en.wikipedia.org/wiki/MapReduce
12. Rizvi SS, Chung TS (2010) Flash memory SSD based DBMS for high performance computing embedded and multimedia systems. In: Proceedings of the 2010 international conference on computer engineering and systems (ICCES), pp 183–188
13. Wikipedia: Cloud database (2013). http://en.wikipedia.org/wiki/Cloud_database
14. Li Jq, Wang Sl, Niu Cl, Liu Jz (2008) Research and application of data mining technique in power plant. In: Proceedings of the 2008 international symposium on computational intelligence and design (ISCID), vol 2, pp 250–253
15. Huang, JA, Vanier G, Valette A, Harrison S, Wehenkel L (2003) Application of data mining techniques for automat settings in emergency control at Hydro-Quebec. In: Proceedings of the 2003 IEEE power engineering society general meeting, vol 4, pp 2037–2044
16. Swartz RA, Lynch JP, Zerbst S, Sweetman B, Rolfes R (2010) Structural monitoring of wind turbines using wireless sensor networks. Smart Struct Syst 6:1–14
17. Ben-Yaacov GZ (1979) Interactive computation and data management for power system studies. Comput J 22:76–79
18. Papadakis M, Hatzjargyriou N, Gazidellis D (1989) Interactive data management system for power system planning studies. IEEE Trans Power Syst 4:329–335
19. GE Power Controls. http://www.gepowercontrols.com
20. Siemens Energy. http://www.energy.siemens.com
21. Simpson, RH (2000) Power system database management. In: Conference record of 2000 annual pulp and paper industry technical conference (PPIC), pp 79–83
22. Wikipedia: Common information model (electricity) (2013) http://en.wikipedia.org/wiki/Common_Information_Model_(electricity)
23. Simmins JJ (2011) The impact of PAP 8 on the Common Information Model (CIM). In: Proceedings of the 2011 IEEE/PES power systems conference and exposition (PSCE), pp 1–2
24. Wikipedia: Generic substation events (2013) http://en.wikipedia.org/wiki/Generic_Substation_Events
25. Wikipedia: Substation configuration language (2013) http://en.wikipedia.org/wiki/Substation_Configuration_Language
26. Morais J, Pires Y, Cardoso C, Klautau A (2009) An overview of data mining techniques applied to power systems. In: Ponce J, Karahoca A (eds) Data mining and knowledge discovery in real life applications. I-Tech education and publishing
27. Martinez C, Huang H, Guttromson R (2005) Archiving and management of power systems data for real-time performance monitoring platform. Tech. rep, Consortium of electric reliability technology solutions

28. Qiu J, Liu J, Hou Y, Zhang J (2011) Use of real-time/historical database in smart grid. In: Proceedings of the 2011 international conference on electric information and control engineering (ICEICE), pp 1883–1886
29. Owoola MA (2004) A generic spatial database schema for a typical electric transmission utility. In: Proceedings of the geospatial information and technology association's 27th Annual Conference (GITA), pp. 1–12
30. Lu B, Song W (2010) Research on heterogeneous data integration for smart grid. In: Proceedings of the 2010 3rd IEEE international conference on computer science and information technology (ICCSIT), vol 3, pp. 52–56
31. SISCO Inc: Integration of substation data. http://cimug.ucaiug.org/KB/Knowledge Base/ Integration of Substation Data ver 06.pdf
32. Zheng L, Chen S, Hu Y, He J (2011) Applications of cloud computing in the smart grid. In: Proceedings of the 2nd international conference on artificial intelligence, management science and electronic commerce (AIMSEC), pp 203–206
33. Rusitschka S, Eger K, Gerdes C (2010) Smart grid data cloud: a model for utilizing cloud computing in the smart grid domain. In: Proceedings of the 1st IEEE international conference on smart grid communications (SmartGridComm), pp 483–488
34. Wikipedia: Outage management system (2013) http://en.wikipedia.org/wiki/ Outage_management_system
35. Awerbuch S, Preston AM (1997) The virtual utility: accounting technology and competitive aspects of the emerging industry. Kluwer Academic Publisher
36. Kaplan SM, Sissine F, Abel A, Wellinghoff J, Kelly SG, Hoecker JJ (2009) Smart grid: modernizing electric power transmission and distribution; energy independence, storage and security; energy independence and security Act of 2007 (EISA); Improving electrical grid efficiency, communication, reliability, and resiliency; integrating new and renewable energy sources. TheCapitol.Net, Inc.
37. Wikipedia: Automatic meter reading (2013) http://en.wikipedia.org/wiki/Automatic_meter_ reading
38. Wikipedia: Advanced metering infrastructure (2013) http://en.wikipedia.org/wiki/Advanced_ Metering_Infrastructure
39. Hart DG (2008) Using AMI to realize the smart grid. In: Proceedings of the conference on power and energy society general meeting - conversion and delivery of electrical energy in the 21st Century, pp 20–24
40. Lui TJ, Stirling W, Marcy HO (2010) Get smart: using demand response with appliances to cut peak energy use, drive energy conservation, enable renewable energy sources and reduce greenhouse-gas emissions. IEEE Power Energy Mag 8:66–78
41. IBM Software Group (2012) Managing big data for smart grids and smart meters. IBM Corporation, Tech. rep
42. Arenas-Martinez M, Herrero-Lopez S, Sanchez A, Williams JR, Roth P, Hofmann P, Zeier A (2010) A comparative study of data storage and processing architectures for the smart grid. In: Proceedings of the 1st IEEE international conference on smart grid communications (SmartGridComm), pp 285–290
43. Fan Z (2011) Distributed demand response and user adaptation in smart grids. In: Proceedings of the 2011 IFIP/IEEE international symposium on integrated network management (IM), pp 726–729
44. Han J, Kamber M, Pei J (2011) Data mining: concepts and techniques. Morgan Kaufmann Publishers
45. Talia D, Trunfio P (2010) How distributed data mining tasks can thrive as knowledge services? Commun ACM 53:132–137
46. Gama J (2010) Knowledge discovery from data streams. Chapman and Hall/CRC
47. Last M, Kandel A, Bunke H (2004) Data mining in time series databases. Word Scientific Press

48. Li D, Aung Z, Williams J, Sanchez A (2012) P3: privacy preservation protocol for appliance control application. In: Proceedings of the 3rd IEEE international conference on smart grid communications (SmartGridComm), pp 294–299
49. Lindell Y, Pinkas B (2000) Privacy preserving data mining. In: Proceedings of the 20th annual international cryptology conference on advances in cryptology (CRYPTO), pp 36–54
50. Kursawe K, Danezis G, Kohlweiss M (2011) Privacy-friendly aggregation for the smart-grid. In: Proceedings of the 11th international symposium on privacy enhancing technologies (PETS), pp 175–191
51. Li DHW, Cheung GHW, Lam JC (2005) Analysis of the operational performance and efficiency characteristic for photovoltaic system in Hong Kong. Energy Convers Manag 46:1107–1118
52. Fan S, Chen L, Lee W (2008) Short-term load forecasting using comprehensive combination based on multi-meteorological information. In: Proceedings of the 2008 IEEE/IAS industrial and commercial power systems technical conference (ICPS), pp 1–7 19
53. Deng J, Jirutitijaroen P (2010) Short-term load forecasting using time series analysis: a case study for Singapore. In: Proceedings of the 2010 IEEE conference on cybernetics and intelligent systems (CIS), pp 231–236
54. Hong T (2010) Short term electric load forecasting. Ph.D. thesis, North Carolina State University, USA
55. Zhang HT, Xu FY, Zhou L (2010) Artificial neural network for load forecasting in smart grid. In: Proceedings of the 2010 international conference on machine learning and cybernetics (ICMLC), vol 6, pp 3200–3205
56. Aung Z, Toukhy M, Williams J, Sanchez A, Herrero S (2012) Towards accurate electricity load forecasting in smart grids. In: Proceedings of the 4th international conference on advances in databases, knowledge, and data applications (DBKDA), pp 51–57
57. Taylor JW (2008) An evaluation of methods for very short term electricity demand forecasting using minute-by-minute British data. Int J Forecast 24:645–658
58. Krishnaswamy S (2012) Energy analytics: when data mining meets the smart grid. http://smartgrid.i2r.a-star.edu.sg/2012/slides/i2r.pdf
59. Ramchurn SD, Vytelingum P, Rogers A, Jennings NR (2012) Putting the "smarts" into the smart grid: a grand challenge for artificial intelligence. Commun ACM 55:86–97
60. Wikipedia: Dissolved gas analysis (2013) http://en.wikipedia.org/wiki/Dissolved_gas_analysis
61. Sharma NK, Tiwari PK, Sood YR (2011) Review of artificial intelligence techniques application to dissolved gas analysis on power transformer. Int J Comput Electr Eng 3:577–582
62. Chen Y, Huang Z, Liu Y, Rice MJ, Jin S (2012) Computational challenges for power system operation. In: Proceedings of the 2012 Hawaii international conference on system sciences (HICSS) pp 2141–2150
63. Zhong W, Sun Y, Xu M, Liu J (2010) State assessment system of power transformer equipments based on data mining and fuzzy theory. In: Proceedings of the 2010 international conference on intelligent computation technology and automation (ICICTA), vol 3, pp 372–375
64. Samantaray SR, El-Arroudi K, Joós G, Kamwa I (2010) A fuzzy rule-based approach for islanding detection in distributed generation. IEEE Trans Power Delivery 25:1427–1433
65. Najy W, Zeineldin H, Alaboudy AK, Woon WL (2011) A Bayesian passive islanding detection method for inverter-based distributed generation using ESPRIT. IEEE Trans Power Delivery 26:2687–2696
66. Calderaro V, Hadjicostis C, Piccolo A, Siano P (2011) Failure identification in smart grids based on Petri Net modeling. IEEE Trans Ind Electron 58:4613–4623
67. Xu L, Chow MY, Taylor LS (2007) Power distribution fault cause identification with imbalanced data using the data mining-based fuzzy classification E-algorithm. IEEE Trans Power Syst 22:164–171

68. Adolf R, Haglin D, Halappanavar M, Chen Y, Huang Z (2011) Techniques for improving filters in power grid contingency analysis. In: Proceedings of the 7th international conference on machine learning and data mining in pattern recognition (MLDM), pp 599–611
69. He H, Starzyk J (2006) A self-organizing learning array system for power quality classification based on wavelet transform. IEEE Trans Power Delivery 21:286–295
70. Hongke H, Linhai Q (2010) Application and research of multidimensional data analysis in power quality. In: Proceedings of the 2010 international conference on computer design and applications (ICCDA), vol 1, pp 390–393
71. Gross P, Boulanger A, Arias M, Waltz D, Long PM, Lawson C, Anderson R, Koenig M, Mastrocinque M, Fairechio W, Johnson JA, Lee S, Doherty F, Kressner A (2006) Predicting electricity distribution feeder failures using machine learning susceptibility analysis. In: Proceedings of the 18th conference on innovative applications of artificial intelligence (IAAI), vol 2, pp 1705–1711
72. Mori H (2006) State-of-the-art overview on data mining in power systems. In: Proceedings of the 2006 IEEE PES power systems conference and exposition (PSCE), pp 33–34
73. Martínez-Álvarez F, Troncoso A, Riquelme JC, Aguilar-Ruiz JS (2011) Energy time series forecasting based on pattern sequence similarity. IEEE Trans Knowl Data Eng 23:1230–1243
74. Neupane B, Perera KS, Aung Z, Woon WL (2012) Artificial neural network-based electricity price forecasting for smart grid deployment. In: Proceedings of the 2012 IEEE international conference on computer systems and industrial informatics (ICCSII), pp 1–6
75. Fernandez I, Borges CE, Penya YK (2011) Efficient building load forecasting. In: Proceedings of the 16th IEEE conference on emerging technologies and factory automation (ETFA), pp 1–8
76. Edwards RE, New J, Parker LE (2012) Predicting future hourly residential electrical consumption: a machine learning case study. Energy Buildings 49:591–603
77. Chicco G, Napoli R, Postolache P, Scutariu M, Toader C (2003) Customer characterization options for improving the tariff offer. IEEE Trans Power Syst 18:381–387
78. Fernandes RAS, Silva IN, Oleskovicz M (2010) Identification of residential load profile in the Smart Grid context. In: Proceedings of the 2010 IEEE power and energy society general meeting, pp 1–6
79. Figueiredo V, Rodrigues F, Vale Z, Gouveia JB (2005) An electric energy consumer characterization framework based on data mining techniques. IEEE Trans Power Syst 20:596–602
80. Li D, Aung Z, Williams J, Sanchez A (2012) Efficient authentication scheme for data aggregation in smart grid with fault tolerance and fault diagnosis. In: Proceedings of the 2012 IEEE PES conference on innovative smart grid technologies (ISGT), pp 1–8
81. Faisal MA, Aung Z, Williams JR, Sanchez A (2012) Securing advanced metering infrastructure using intrusion detection system with data stream mining. In: Proceedings of the 2012 Pacific Asia workshop on intelligence and security informatics (PAISI), pp 96–111
82. Fatemieh O, Chandra R, Gunter CA (2010) Low cost and secure smart meter communications using the TV white spaces. In: Proceedings of the 2010 IEEE international symposium on resilient control systems (ISRCS), pp 1–6

Chapter 8
Securing the Smart Grid: A Machine Learning Approach

A. B. M. Shawkat Ali, Salahuddin Azad and Tanzim Khorshed

Abstract The demand of electricity is increasing in parallel with the growth of the world population. The existing power grid, which is over 100 years old, is facing many challenges to facilitate the continuous flow of electricity from large power plants to the consumers. To overcome these challenges, the power industry has warmly accepted the new concept *smart grid* which has been initiated by the engineers. This movement will be more beneficial and sustainable to the extent if we can offer a secure smart grid. *Machine learning*, representing a comparatively new era of Information Technology, can make smart grid really secure. This chapter provides an overview of the smart grid and a practical demonstration of maintaining the security of smart grid by incorporating machine learning.

8.1 Introduction

Due to growing concern over massive carbon emission and consequently, rapid climate change, there has been a global movement towards the discovery of clean and renewable energy sources. Traditional energy sources, like fossil fuel, are non-renewable in the sense that they take millions of years to form. Due to huge energy demand worldwide, these traditional energy sources are being exhausted more quickly than the ones being created. The most significant advantage of renewable energy sources are that they are never going to run out as they are refilled naturally. Another advantage is that almost all of the renewable energy sources are relatively clean and cause minor or no pollution to the environment. The most prominent renewable energy source is the *solar* power. The other significant sources are *wind power*, *ocean waves*, *hydropower*, *biomass power* and *geothermal energy*.

A. B. M. S. Ali (✉) · S. Azad · T. Khorshed
Central Queensland University, Rockhampton, QLD 4702, Australia
e-mail: s.ali@cqu.edu.au

A. B. M. S. Ali (ed.), *Smart Grids*, Green Energy and Technology,
DOI: 10.1007/978-1-4471-5210-1_8, © Springer-Verlag London 2013

Due to the technological advancements, increased automation and ever increasing number of consumers, the hunger for energy is likely to aggravate. As the supply from traditional energy sources is running low, there is a tremendous urge globally to become energy efficient. The practice of being energy efficient not only saves energy but also helps to reduce the amount of carbon discharged in the environment during the generation of power. Energy efficiency can be ensured through minimizing of the loss during generation, transmission and distribution phase on the power production side. On the consumer side, energy efficiency can be guaranteed through the design and use of devices that would consume minimum energy during their operations. The consumers also have the responsibility to use power in a sensible way i.e., they should not keep the devices turned on unnecessarily. Alternatively, the device itself should detect when it is not being used and should switch to sleep mode as soon as possible.

The function of an electric power grid is to carry out mass transfer of electric energy at high voltage from power plants, where power is generated, to the substations located near the customer base. In the substations, the electricity is stepped down in voltage and passed through distribution wiring to the consumer site where the power is further stepped down to service voltage. The major problem with tradition electricity distribution systems are that the energy produced cannot be stored and therefore, should be generated as required. When the supply and demand is not in equilibrium, the generation units and transmission network can be shut down causing blackouts. The introduction of alternative power sources, which are intermittent in nature, has made the stable power supply more difficult. Traditional distribution systems are vulnerable to security threats either from energy suppliers or cyber attacks.

The concept of *smart grid* emerged as result of the desperate attempts to make the power grid stable, reliable, efficient and secure. Smart grid attempts predict the usage pattern of the customers and respond intelligently in order to provide reliable, sustainable, and cost-effective services. One significant aspect of the smart grid is that it can schedule and control the load to effectively shift the usage to off-peak hours and reduce the peak demand, which is known as *demand management*. Dynamic pricing is another mechanism of the smart grid to facilitate demand management, which motivates the consumers by increasing the electricity price during the high demand period and reducing the electricity price during the low demand periods. According to the Energy Future Coalition's Smart Grid Working Group, a smart gird should incorporate the following functionalities [1]:

- Would give consumers control over their usage.
- Increase the efficiency and be more economical.
- Should be self-healing and more secure both from physical and cyber attacks.
- Should be able to integrate alternate energy sources like solar cell or wind power.

Smart grid is seen as a way to allow consumers to take part in optimizing the operations by providing them greater information and options. Figure 8.1 describes the operation of self-healing smart grid.

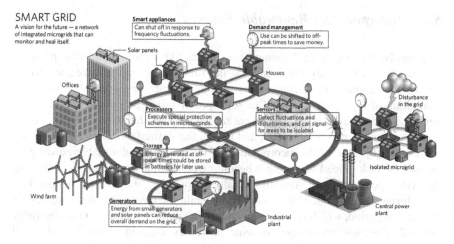

Fig. 8.1 A smart self-healing grid system [40]

8.2 Smart Power Generation

The main challenge in the current grid is that electricity demand fluctuates throughout the day and also varies from one season to another [2]. Therefore, one of the key factors to maintain reliability and stability in power supply is load balancing. The load balancing task is performed by the transmission operators by matching the power output with the load. In the traditional power grid, combination of three types of power plants—*baseload*, *load follower* and *peaking* are installed to mitigate the balancing task, while maintaining the economy.

Baseload power plants [3] are the power generation facility that produces the minimum amount of energy that is continuously required by a region. Baseload plants operate at the maximum capacity and they typically generate power round the clock and are only shut down for repair and maintenance. Baseload plants are, in general, based on nuclear energy, coal or geothermal energy. *Load follower* power plants [3] generate power only during period of high demand, for example, during daytime or early evening. Gas turbine combined cycle and hydroelectric power plants are typical load followers. *Peaking* power plants [3] generate power when the power demand rises for a short period. Single cycle gas turbines or gas engines are usually deployed as peaking power plants. The baseload plants have the lowest fuel cost followed by the load follower plants and the peaking plants have the highest fuel cost. However, the peaking plants have the shortest start-up time, which means they can start within a very short time during the period of peak demand.

8.2.1 Incorporating Renewable Sources

The supply of power from the renewable sources like solar cell and wind power is not steady and precisely predictable as the time of the day and weather events dictate the output from these sources. It was initially envisaged that statistical smoothing of high peaks of wind or solar power might be possible by interconnecting adjacent areas [4, 5]. But recent research shows that wind and solar power patterns match across a large area and hence, the desired smoothing affect is not achievable [6]. Due to the incorporation of intermittent and variable power generators, the task of load balancing is becoming increasingly complex. For example, the wind speed in South Australia usually drops when the ambient temperature rises during the day [6]. The demand rises, while the electricity production drops during that period, which results in a wind output pattern contrary to the electricity demand. The planned increase in wind power by 2020 in South Australia will severely impact the balance dynamics.

Intermittent power output from renewable sources often causes temporary high-peaks and extended period of zero output. The instantaneous output from the renewable sources is sometimes so huge that it can even drive baseload plants out of production. As the baseload plants require high investment, if the baseload plants remain shut down and operate at reduced load, the capital cost per kWh will rise drastically due to high investment behind the plants. It would be smarter if energy from renewable sources can be stored and used later to flatten the peaks and valleys.

8.2.2 Energy Storage Technologies

Flywheels can be accelerated by electricity and this energy will be preserved as rotation energy. The flywheels can drive generators to cover up sudden shortage of power. The rotational energy will eventually be lost due to friction in the bearing. Putting the rotor inside a vacuum chamber and using magnetic bearing can minimize the energy loss. Electric vehicle batteries may also be used as virtual energy storage which can be charged during off-peak hours and drained off during peak hours. The above mentioned peak-shaving technologies can only provide backup for a short period of time. Storage of energy as thermal energy can be an effective way to balance the demand and supply. Molten salt is heated using solar energy and sent to hot storage tank. This hot salt is later used to produce superheated steam for electricity generation. Surplus renewable energy can also be used to compress air to a high pressure and stored for a long time. The pressurized air can later be used as combustion air in the peaking power plants. Stored natural gas or biogas provides higher energy density and hence, better alternatives to compressed air. Pumped hydro-storage system can pump water from low-altitude reservoir to a high-altitude reservoir. During the period of high demand, this water is passed through the turbines to produce electricity. However, most of the solutions are still more expensive than conventional peaking power plants. Also they are unable to provide backup for extended period of time with zero output from wind or solar power plants. So far, pumped hydro-storage has been proved to be the most economical alternative to the conventional peaking plants. However, pumped hydro-storage system requires huge unoccupied land for the construction of reservoirs.

Recently, hydrogen is emerging as a storage medium. Hydrogen can be produced using electricity or heat and stored in compressed or liquid form. The hydrogen is later converted back to electricity by a combustion engine. An alternative way is to use the hydrogen as a fuel for fuel cell that can convert chemical energy into electricity.

8.2.3 Mitigation of Peak Demand

Since air conditioning is one of the key factors for peak demands, excess energy can be used to produce ice during night time and can be used for air conditioning during daytime. Optimum insulation of buildings is generally the cheapest and most effective solution for reducing the seasonal peaks. Export and import of electricity from neighboring countries can ease the gap between production and demand to a large extent. Cogeneration units use waste heat produced in the power plants for heating purposes either utilizing the heat near the power plants or through district heating. The fuel utilization efficiency of cogeneration units is higher than producing electricity and heat separately [2]. District heating is one of the cheapest methods for reducing carbon footprint.

Demand management rewards the customers for using less energy during the peak time and moves the energy usage to off-peak time so as to shave the peaks and to fill the valleys. Price-based demand management charges the customers at time varying rates that correspond to the value and cost of electricity at different times. Incentive based demand response pay for reducing their loads as requested by the utility provider. Demand management also increases the load during the time of high supply or low demand. The Queensland government is planning to install devices into some household appliances such as air conditioners, pool pumps and hot water systems. The devices would enable the utility companies to cycle the use of these devices during the peak hours [7].

8.2.4 Forecasting Renewable Energy Supply

Wind and solar forecasting technology makes renewable energy supply more reliable and cost-effective as it provides more time to the grid operators to plan in advance for a backup energy supply when the renewable energy source is expected to produce less energy than required. For example, grid operators will not carry much spinning reserve if they know ahead that a sudden drop in renewable is less likely. The objective of wind power prediction is to predict the wind speed and direction, while the objective of solar power prediction is to predict the solar irradiation. Forecasting may be done with a numerical weather prediction model or statistical analysis of local measurements [8]. The first method uses models on meteorological measurements and observations coming from satellites or weather stations. As this is a pure analytical method, no historical data is necessary but the computational complexity is huge. The second method based is on the relationship between the historical data and forecast variables. Since it learns from previous experience, the statistical model can account for the local terrain and other details that can't realistically be represented in the numerical weather prediction model [9]. For the same reason, the statistical model tends to predict typical weather events rather than exceptional events [10].

8.3 Smart Grid Security Issues

Since the smart grid technology is complex which includes generation, transmission, and distribution and hence, there are manifold opportunities for the attackers to disrupt the system. It is a common conception that threats solely comes from hackers or other individuals or groups with malicious intent. In fact, staff and other insiders can also pose a risk as they have authorized access to one or more components of the system. Insiders know about sensitive information of the system such as passwords stored in the database, cryptographic keys and others security mechanisms that could be utilized to organize an attack. However, not all security infringements are malicious; many of them originate from accidental mis-configurations, failure to follow procedures and other oversights.

Smart grid threats can be classified into three broad groups: (1) system level threats that attempt to take down the grid; (2) attempts to steal electrical service; and (3) attempts to compromise the confidentiality of data on the system.

8.3.1 System Level Threats

System level threats attempts to take down part or the entire smart grid. For example, entities or individuals with malicious intent could attempt to change programmed instructions in the meter, change alarm thresholds or issue unauthorized commands to meters or other control device on the grid. This type of actions could result in damage to equipment, premature shutdown of power or processes or even disabling of control equipment. The following are among the system level threats commonly encountered.

8.3.2 Radio Subversion or Takeover

This threat aims to capture one or more radios or the RF communication modules in the meters so they act on behalf of the attackers. The common form of this attack is firmware replacement. Attackers attempt to insert modified firmware into a device or attempt to spread compromised firmware to numerous devices.

8.3.3 Network Barge-in by Strangers

This threat is characterized by attempt of stranger radios to join the RF radio network and/or preventing the communication modules to communicating properly. For example, an attacker may try to use the communication module to piggyback unauthorized traffic through the network or use a stranger radio to intercept or relay traffic. Moreover, an attacker may attempt to modify a radio or communications modules' credentials to assume a different role. Since the interface between the radio and microcontroller is frequently not encrypted, an attacker can participate in the communication as a legitimate device by manipulating the trust relationship.

8.3.4 Denial of Service

This threat results in part or the entire network becoming unresponsive to service request. This attack can take place in form of resource exhaustion or jamming. In routing black holes attack, a node is hacked so that it's advertised as the shortest path to everywhere, resulting in all traffic being diverted to it. RF spectrum jamming attack prevents signal from being received. In jabbering attack, a legitimate node is co-opted to send so much traffic that other nodes can't communicate; Kill packets are protocol packets that cause radios to crash or to become unreachable via the RF field. Stack smashing, a method of subverting or crashing a device's operating system or applications by overloading memory buffers so that data is exposed, lost or corrupted.

   The core elements of the network—routers and switches, if not safeguarded properly, can be compromised and make smart grid vulnerable. For example, routers can be shipped with factory default password and remote access such as Telnet or HTTP services turned on. Network administrators knowingly or unknowingly leaving these default settings unchanged can create an entry point into the system for the intruders. In case the devices are compromised, they can be utilized to disrupt grid operations through denial-of-service (DoS) attack. In the worst case, they can be used to take control of more critical control systems.

8.3.5 *Malicious Code*

Software update feature allows the devices to check for download and install the software packages and patches. If the update is not from trusted vendors, malicious codes such as Trojan horses, or malicious worms can make a way into the system. It is a common practice to write the downloaded software into flash. The code is only executed if it passes the authentication, while the unauthenticated code remains in the flash. This could be executed through code exploit or glitch hardware. The local interfaces are often not secured and create an easy entry point [11].

8.3.6 *Glitching*

No two transistors in the system are the same due to their locations, tolerances and I/O factors. A glitch forces transistors to operate when they shouldn't [12]. The glitch can be injected through power supply, clock signal, and electromagnetic radiation.

8.4 System Level Theft of Service

Theft of service attack consumes service from the utility provider without paying the revenue to the provider. For example, individual meters or a group of meters can be subverted to misreport to the customer, the amount of service provided or the cost of service provided (changing from a higher-priced one to lower-priced one). The following are theft of service threats commonly encountered [13]:

8.4.1 *Cloning*

A perpetrator commits this attack by replacing a meter or radio ID with a duplicate one designed to report zero usage. As a result, the utility providers receive no revenue for the service it provides.

8.4.2 *Migration*

This attack aims at reducing the reported usage and associated bills by swapping a meter (or communication module) from a location reporting high usage to with a meter (or module) from a location reporting low usage.

8.4.3 Meter/Communication Module Interface Intrusion

The communication module connects to meter through a serial port. A perpetrator can disconnect the communication module from the meter or can break into communication module so that it doesn't report any usage information or incorrect usage information.

8.5 Breach of Privacy or Confidentiality

This sort of attack may render personally identifiable information being exposed. Common privacy and confidentiality attacks include:

8.5.1 Meter Compromise

A meter can be broken into to retrieve personal information.

8.5.2 RF Interception

A perpetrator can intercept the packets by passive eavesdropping on the radio network.

8.5.3 Forwarding Point Compromise

If a node on the network is compromised, it can be used to forward traffic to some unauthorized individual or group.

8.5.4 Backbone Network Interception

Packets can be captured as it passes through the backbone IP network.

8.5.5 Bus Sniffing

The interface between the microcontroller and the radio is not often encrypted which can attract bus sniffing. It may be possible for an eavesdropper to capture radio configuration information, cryptographic keys, and network authentication credentials.

8.5.6 Key Compromise

The secret key can be revealed by monitoring the power consumption of the device [11]. Transistors draw more current when switching occurs. Since processing is deterministic and repeatable, each operation in the device leaves an EM signature which can be detected by sampling the current consumed by that device. This is known as *power analysis attack*. *Timing attack* can decode the entire secret key by examining the variation in time a cryptographic operation takes. The secret key can also be inferred from the analysis of electromagnetic radiation emitted from a device. This is known as *electromagnetic attack*.

Using the same symmetric key for encryption across the system can make the system seriously vulnerable. If one of the devices in the system is compromised, the whole system is system is eventually compromised.

8.6 Threat Mitigation

Threat mitigation strategies in smart grid can be broadly classified into four main categories—*physical security*, *privacy and security of data*, *authorization and access control*, and *securing network devices and systems*. Moreover, a number of general strategies are needed to tighten the overall security of the system as discussed in Sect. 8.6.5.

8.6.1 Physical Security

When it comes to physical security, the SCADA network always found to have poorly protected. The primary security measure to secure a smart grid would be to keep the intruders off the premises. This could be achieved through video surveillance, cameras, electronic access control, and emergency response [14].

8.6.2 Privacy and Security of Data

Since different entities are involved in the smart grid, it is crucial to protect the data as it is stored and transmitted. The following measures should be implemented to protect the data in the smart grid [14]:

- Implement firewall functionality to impose access policies between different segments of the smart grid.
- Deployment of VPN architecture that encrypts data to ensure secure and confidential data transmission.
- Leverage encryption and data storage security capabilities to protect critical assets on servers and endpoints.
- Provide granular access to sensitive data at the application level.
- Provide ubiquitous and consistent security measures across wired and wireless security connections.

Moreover, care must be taken in the following cases [14]:

- Avoid using insecure remote management and communication protocols.
- Avoid failed authentication account lockout and logging weaknesses.
- Implement certificate revocation list checking practices.

8.6.3 Authentication and Access Control

The smart grid is accessed by various user groups such as employees, contractors or even customers. Access to these user groups, be it local or remote, must be granular and authorization should be granted to 'need to know' assets.

Identity must be verified through strong authentication such as multifactor authentication. Passwords should strong, attempts must be logged and unauthorized attempts also should be logged. Moreover, all access points should be hardened to avoid any loophole in the system. Only ports and services required for normal operation should be enabled.

Communication modules in the meters should use cryptographic keys and digital signatures to confirm that the firmware is from a genuine source and not been tempered. The secret keys must be stored securely and properly protected. One way to achieve this is requiring a password to control the use of the secret key. Blinding the cryptographic operations so that the timing of the operation does not depend on the key can protect against timing attack.

8.6.4 Securing Networked Devices and Systems

Meters should be equipped with temper detection mechanisms. Local temper detection systems are fitted with physical indicators that the meter has been tempered. In remote temper detection mechanism, the meter notifies the head-end about the tempering. System integrity protection system allows the meter to protect its integrity by self-erasure of keys and firm-ware.

To prevent bus sniffing or bus injection, the microcontroller and the radio are placed on the same chip in the meter. However, some devices still have bus sniffing turned on for debugging purposes.

A unique key should be assigned for per device and per use rather than using a single key for all devices across the system. Caution must be taken when cryptographic errors take as it may be an indication of an attack (such as glitching). Care must be taken so that local interfaces are disabled by software.

8.6.5 Maintaining Overall Security

In spite of having discrete functional zones and clear segmentation, it is hard to predict what form of attack would take place. A comprehensive defense strategy is required which would largely cover all sorts of threats and vulnerabilities a smart grid can encounter. An effective, layered defense mechanism should be put in place, which would implement the security principles across the whole infrastructure. The security of the system should be reviewed by the third parties.

System and software development lifecycle implement security all layers (design, coding, testing, deployment and maintenance). Security should be enforced by default instead of making it the responsibility of the end user.

An intrusion detection system (IDS) and/or an intrusion prevention system (IPS) should be implemented to identify external threats trying to penetrate the system and stop any attempts at internal propagation. Host protection mechanism should be deployed to protect critical client systems such as clients, servers and endpoints. Antivirus software and host-based IPSs should be kept up-to-date with latest threat intelligence and signature update.

8.7 An Intelligent DoS Attack Prevention Mechanism

As mentioned earlier in this chapter, the DoS attack is basically an attempt to make a smart grid resource unavailable to its existing users. Although the methods to carry out, motives for, and targets of a DoS attack may differ, it commonly consists of the efforts of an individual or a group of people to temporarily or indefinitely interrupt or suspend the services of smart grid. Distributed denial-of-service

(DDoS) attack is the distributed version of DoS attack, where the attacker uses multiple computers to launch the attacks. Modern DDoS attacks use new techniques to exploit areas where the traditional security solutions are lacking. These attacks can cause severe network downtime to the systems that heavily depend on networks and servers. The number of DDoS attacks remained steady in the recent times but complex multi-vector attacks are becoming more common.

We should not assume that all disruptions to service in the smart grid are the result of a DoS attack. Sometime, the grid may have technical problems with a particular network, or the system administrators may be performing maintenance works. However, the following symptoms *could* indicate a DoS or DDoS attack in the smart grid [15]:

- unusually slow grid network performance
- unavailability for the administrator/customer to get access to a particular or any Website
- dramatic increase in the amount of spam e-mails in the user account

Until today there are no effective ways to prevent being the victim of a DoS or DDoS attack, but there are steps we can take to reduce the likelihood that an attacker will use smart grid network computer to attack user computers. The steps are as follows:

- Install and keep updated anti-virus and operating system software.
- Install a firewall, and configure it to restrict traffic coming into and leaving your computer.
- Following proper security practices.
- Applying email filters may help you manage unwanted traffic.

This chapter demonstrates an intelligent DoS prevention mechanism using machine learning techniques to reduce the likelihood of this sort of attack. Machine learning techniques have been proved to be a very useful tool to prevent a DOS attack [16]. If there is a known type of attack, machine learning can take proactive action to address the issue, and at the same time, notify systems/security administrators as well as the data owner. If an unknown type of attack happens, machine learning will still be able to detect it as an attack from the performance variations from the standard usage, and can notify the designated person with the closest type attack known to its database. That would make the security administrator's job easier and help the administrator fight against unknown types of attacks. In the previous research experiments, the authors successfully identified Cloud insiders activities [17] and DoS/DDoS attacks [18] using machine learning techniques. The previous research found that rule based technique C4.5 is an efficient technique to identify this sort of attacks. The authors validated the performance of different machine learning techniques with the rigorous testing of tenfold cross validation. The experimental outcome demonstrated that C4.5 algorithm not only performs better than other techniques, but also the level of performance is of acceptable standard. The other algorithms tested were Naive

Bayes, Multilayer Perceptron, SVM and PART [17, 18]. This chapter investigates the two groups of machine learning algorithms: *Statistical based* and *Rule based* algorithms to stop DoS attack in the Smart Grid Environment.

8.8 Data Collection

As discussed in the previous section, DoS and DDoS attacks cause significant changes in system performances, and machine learning can easily identify when noteworthy change in system performance occurs. Also [17, 19] suggest that some activities carried out in modern day systems may cause major changes in the performance graph and look very similar to some of the DoS/DDoS attacks in naked eyes. In this situation, machine learning can play a significant role by distinguishing an attack from an activity as it can work with multiple system performance parameters simultaneously that is not possible by human being.

In this section, the pictures of the performance charts that are taken from the modern day server during the cyber attack are presented. The performance plot generated by the data collection spreadsheet during the attack and the similarity between these two indicates are also presented. The primary intention was to train machine with some of the well known DoS/DDoS attacks types and also to train with some normal activities that are carried out in modern network environments every day. So that the machine can distinguish between these two types of activities and can detect unknown type of attacks which are not recorded in the database as an attack or an activity. All these cyber attacks were generated in an experimental environment with some real attack tools. It is to be noted that the performance data of 20 different parameters of System, CPU, Memory and network were collected. In this chapter, only performance charts and plots of those that show significant changes during an attack are included, however, to refine the data using machine learning, all 20 parameters at the same time, irrespective of whether they made any noteworthy dissimilarity or not, are included.

8.8.1 Similarity Between Attacks and Other Activities

Accurate data collection is very important to achieve correct results from machine learning, and also, to make a distinction between an attack and a non-attack activity. To give example of similarity between an attack and a non-attack activity, at first three charts of three different activities are presented. Figure 8.2 presents the disk performance chart for the period of cloning a Virtual Machine (VM) that looks exceptionally similar to disk performance chart at some point in DoS attack using RDoS. Figure 8.3 represents network performance chart while taking the snapshot of a VM over the network and the chart looks incredibly alike to the network performance chart during SYN flood attack. Figure 8.4 shows network

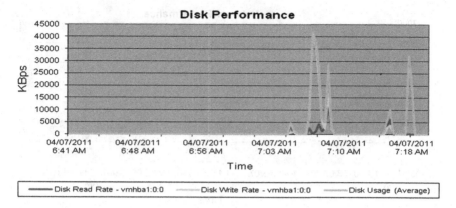

Fig. 8.2 Disk performance chart during the cloning of a VM; it looks incredibly similar to the disk performance chart during DoS attack using RDoS

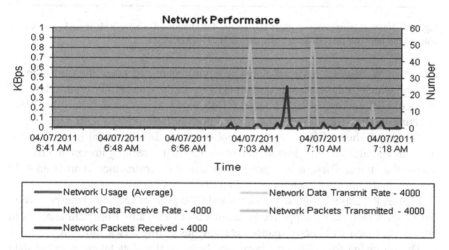

Fig. 8.3 Network performance chart while taking a snapshot of a VM over the network; it looks exceptionally identical to the network performance chart during SYN flood attack

performance chart during installation of new VM and it looks very comparable to the network performance chart during HTTP-DoS attacks.

8.8.2 DoS Using Real-time Disk Operating System

Real-time disk operating system (RDoS) by Rixer [20] is one of the most easily available DDoS attack tool for Web attack. This tool together with a port scanner can be very useful DDoS attacking tools for Web resources. In our experiment the authors only used RDoS and did not use any port scanner as the authors created

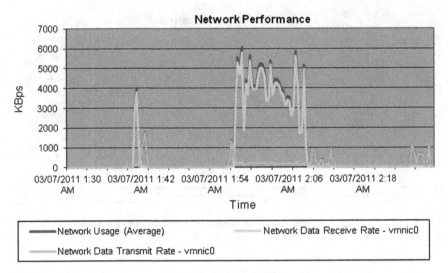

Fig. 8.4 Network performance chart during the installation of a new VM; it looks especially alike to the network performance chart during HTTP-DoS attack

their own Website on a virtual network environment and knew the port number already, which in this case was default HTTP port 80. The HTTP server's Internet Protocol (IP) address is 10.1.1.1 and RDoS tool was executed from other VMs selecting victim's IP address 10.1.1.1 and port 80 from 4:30 to 4:40 a.m.

Figure 8.5 shows RDoS by Rixer tool operation in the virtual network environment and also victim's system performance chart screening important changes during the attack. There is a notable change in the performance chart from 4:30 a.m. onwards since the authors ran this tool.

Figures 8.6 and 8.7 represent System and CPU performance charts respectively, during the attack which happened between 4:30 to 4:40 a.m. Important changes in both System and CPU performances are obvious during attack moment.

The following four diagrams presented here are the plots taken from the data collection spreadsheet. These are exactly the same as the hypervisor performance charts during the time of attack, which indicates how precisely the performance data were collected. Figures 8.8 and 8.9 show the performance plots of CPU and Disk respectively, generated by the data collection spreadsheet.

8.8.3 HTTP-DoS Attack Using Low Orbit Ion Cannon

Low Orbit Ion Cannon (LOIC) is an open source network for stress testing and DoS/DDoS attack application [21, 22]. An attacker can flood TCP/UDP packets with the intention of disrupting the service of a particular host. On December 2010, BBC report entitled "Anonymous Wikileaks supporters explain Web attacks"

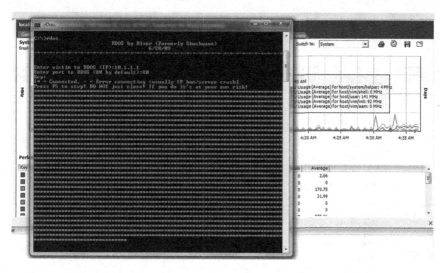

Fig. 8.5 DoS attack using RDoS; the system performance chart screening significant changes during the attack [18]

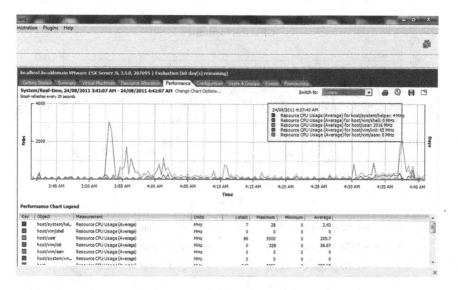

Fig. 8.6 System performance chart during RDoS attack that happened between 4:30 to 4:40 a.m

quoted security experts that well-written firewall rules can filter out most traffic from harmful DDoS attacks by LOIC [23]. However, in the previous research by the authors discovered that these corporate firewalls are not very effective if the attacker resides or shares the same physical hardware from same service provider [24]. For that reason, here, the authors attacked a specific VM from other VMs that is sharing the same physical resources. The HTTP-DoS attack started on victim

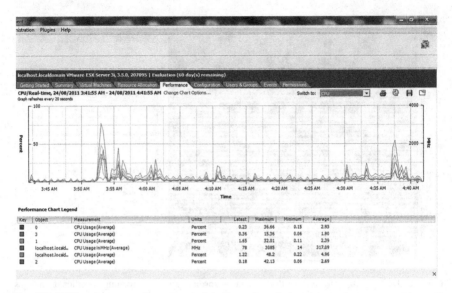

Fig. 8.7 CPU performance chart generated in hypervisor during the attack

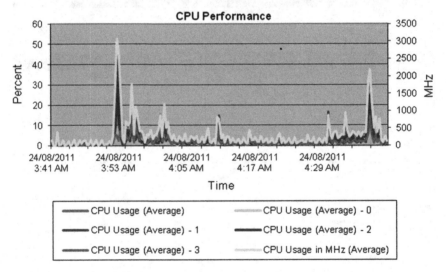

Fig. 8.8 CPU performance plot generated by data collected for the experiment; It has similarity with the one generated automatically by the hypervisor

(IP 10.1.1.1) using LOIC at 6:08 a.m. and ended at 6:15 a.m. Figure 8.10 shows LOIC running from attacker VM with target IP 10.1.1.1.

Figures 8.11, 8.12 and 8.13 present the performance plots of CPU, network and system generated by the data collection spreadsheet that were collected from the

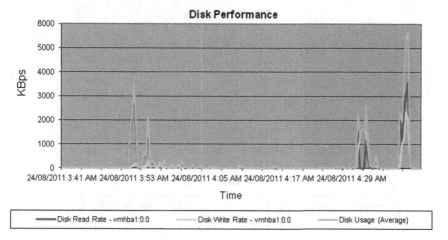

Fig. 8.9 Performance plot disk performance from the data collection spreadsheet

Fig. 8.10 HTTP-DoS attack using LOIC running, also showing CPU performance chart generated in hypervisor [18]

Virtual machine manager (VMM) during the attack. Sudden increase in the performance was noticed during the time of attack (start 6:08 a.m. and ended at 6:15 a.m.).

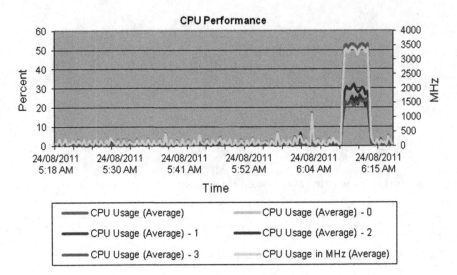

Fig. 8.11 CPU performance plot generated from the data collection during HTTP-DoS attack [24]

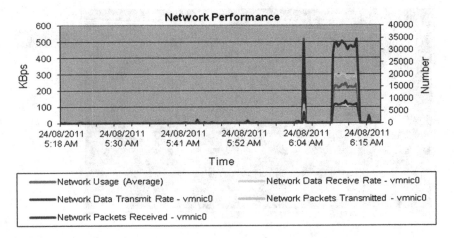

Fig. 8.12 Network performance plot generated from the collected data [24]

8.8.4 Ping Flood Attack

Ping flood is another kind of DoS attack where the attacker crushes the victim with Internet Control message protocol (ICMP) Echo Request (ping) packets. This method could be very successful when sending packets quickly without waiting for a response from the victim. If ICMP service is not disabled by the target host, it will flood the target host with large data segments [25, 26]. However, in our study and from work experience, we found organizations usually disable ICMP requests

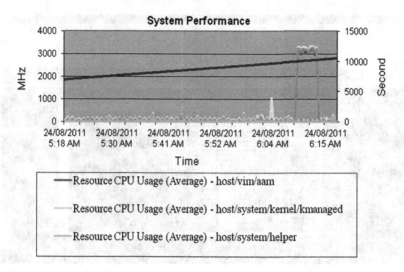

Fig. 8.13 System performance plot generated from the data collected

at firewall or in the router so that it can stop ICMP requests from external networks, traditionally they keep ICMP open on hosts in their own internal networks
so that they can do network diagnostics. Our concern for modern VMs is that an
attacker could be residing on the same physical hardware or somehow can manage
to hack into another low secured VM that is residing on same internal virtual
network and, carry this kind of attack to a target VM.

A certain kind of ping flood attack in the past was named "ping of death" where
an attacker deliberately used to send packets larger than the 65,536 bytes, many
computer systems were not able to handle a ping packet larger than this maximum
IPv4 packet size [27]. So, in our experiment, we send ICMP packets from each
attacker VM slightly lower than that so that the attacker VM itself does not get
overwhelmed. We ran "ping 10.1.1.1 –t –l 65000" command from each attacker
VM. Here –t was used for repeated sending of echo messages and –l indicates the
size of packet to be sent, in this case it was 65,000 bytes from attacker VM1 (we
named it win7_1 as shown in Fig. 8.14).

Figure 8.14 shows hypervisor console running ping flood attack from attacker
VM to Victim VM. While Fig. 8.15 represents network performance chart from
VMM in the instance of attack and Fig. 8.16 is the performance plot of network
generated by our data collection spreadsheet.

8.8.5 SYN Flood Attack Using Engage Packet Builder

A SYN flood attack is also another kind of DoS attack where a network becomes
overwhelmed by a series of SYN requests to a target's system [28]. Engage Packet

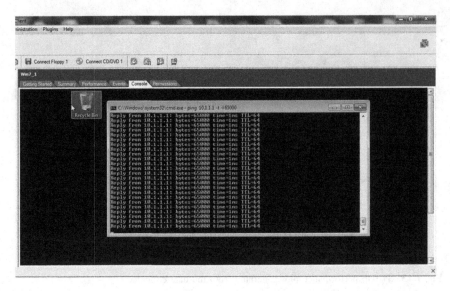

Fig. 8.14 Snapshot of the hypervisor console running ping flood attack from attacker VM to Victim VM

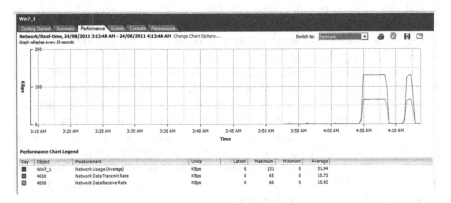

Fig. 8.15 Network performance chart generated from the hypervisor

Builder [29] is a powerful and scriptable packet builder with capability of packet injection starting from link layer (MAC address spoofing), it can also generate SYN-Floods by building "strange" packets [29]. We used Engage Packet builder to execute SYN flood attack twice, at 5:13 and 5.17 a.m. Figure 8.17 shows Engage Packet Builder running from attacker VM (IP 10.1.1.10) with target IP 10.1.1.1 and Fig. 8.18 shows a performance plot of the network generated from the data which was collected from the VMM during the attack.

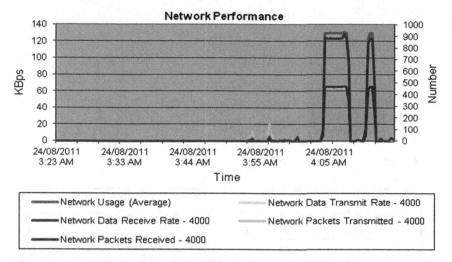

Fig. 8.16 Network performance plot generated from collected data

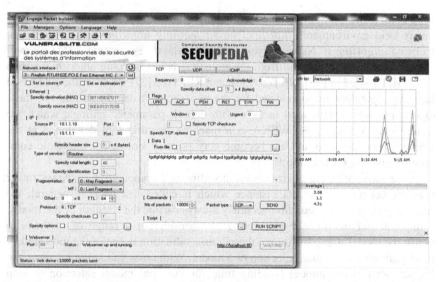

Fig. 8.17 Snapshot of SYN flood attack using engage packet builder and network performance monitoring from the hypervisor during the attack

8.9 Experimental Outcome

The data were collected from a real life environment to measure the strength of the machine learning modeling performance to prevent any DoS attack in the Smart Grid environment. The total numbers of instances in the dataset are 536 and the number of attributes is 21. All the attributes are numeric in the dataset. The authors

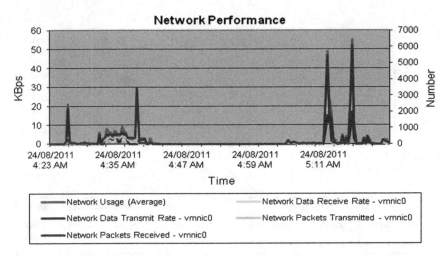

Fig. 8.18 Network performance plot generated from the data collected during the SYN flood attacks

have chosen two sets of machine learning algorithms to prevent DoS attack in the Smart Grid environment: *statistical based learning algorithms* and *rule based learning algorithms*. Naive Bayes [30], Multilayer Perceptron [31], Support Vector Machine [32, 33] are from the statistical based learning algorithms group and PART [34], J48 [35], NBTree [36], and REPTree [37] are from the rule based learning algorithm group selected for the experimental demonstration. All these algorithms are implemented in Java with default parameter settings, which are available in WEKA [38]. WEKA is a machine learning tool developed at the University of Waikato and has become very popular among the academic community working on learning theory.

In the data modeling environment, modeling is a comparative easy task rather than predict an unseen data instance which is called test data. In general, the percentage of model prediction accuracy always carries the strength of the model. Computational intelligence researchers are considering a range of methods to verify the model strength. Among these, cross validation is one of the most widely used methods for the final selection of a model. With cross validation measure, the kappa statistics and model building time for the final model selection to stop attacker in the smart grid network was also chosen. In the following, the tenfold Cross Validation method [38] is explained.

The steps of tenfold cross validation procedure are as follows:

- First, use a random sampling procedure to split the entire training set into 10 sub-samples. Let's call these samples S1, S2, S3 and so on, until we get to S10.
- As a first step, remove one sample set, say 10 (S10), from the training set.
- Train the machine learning algorithm using data from S1 to S9.

- Once the machine has built a model based on data from S1 to S9, it sees how accurately the model predicts the unseen data of S10. Error rates are stored by the system.
- Once the accuracy of predicting the values in S10 is tested, S10 is put back into the training set.
- For the next step, we remove sample set S9 from the training set.
- Re-train the machine learning algorithm, this time using data from S1 to S8 and S10 (i.e., leave out S9).
- Once the machine has built a model based in the training set as described in Step 7, it evaluates how accurately it can predict values in the new test set (i.e., S9). Error rates are stored by the system.
- Put S9 back into the training set.
- Now, remove S8 from the training set, and repeat the testing procedure.

At the end of the sequence, the 10 results from the folds can be averaged to produce a single estimation of the model's predictive performance. The main advantage of the tenfold cross validation method is that all observations are used for both training and validation, and each observation is used for validation exactly once. This leads to a more accurate measure of how efficiently the algorithm has "learned" a concept, based on the training set data. Thus, a final model is setup to predict the upcoming new sample.

Experimentally, the accuracy as the overall number of correct classifications averaged across all tenfolds is estimated. Let D_i be the test set that includes sample $v_i = \langle \mathbf{x}_i, y_i \rangle$ and the cross validation accuracy estimation is defined as:

$$\text{Acc}_{cv} = \frac{1}{nf} \sum_{i=1}^{f} \delta\big(\Im\big(D_{(i)}, \mathbf{x}_i\big), y_i\big),$$

where f is the number of folds and n is the number of labeled instances in the fold. Kappa is a chance-corrected measure of agreement between two raters, each of whom independently classifies each of a sample of subjects into one of a set of mutually exclusive and exhaustive categories [39]. It is computed as

$$K = \frac{p_o - p_e}{1 - p_e},$$

where $p_o = \sum_{i=1}^{k} p_{ii}$, $p_e = \sum_{i=1}^{k} p_i.p_{.i}$, and $p =$ the proportion of ratings by two raters on a scale having k categories.

Both statistical and rule based learning algorithms' performances have been summarized in Figs. 8.19, 8.20, 8.21, 8.22, 8.23 and 8.24. Figure 8.19 shows Multilayer Perceptron algorithm as is the best choice for preventing the DoS attach in the Smart Grid. In the Kappa statistics measure of Fig. 8.20, Multilayer Perceptron is again shown as the most superior. However, in terms of computational complexity measure, Multilayer Perceptron was the last choice as shown in Fig. 8.21. Naïve Bayes is a comparatively faster algorithm among the statistical learning algorithms. Figure 8.22 shows that PART algorithm is the best choice

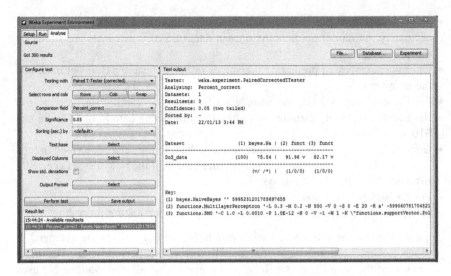

Fig. 8.19 DoS attack classification accuracies of statistical based learning algorithms

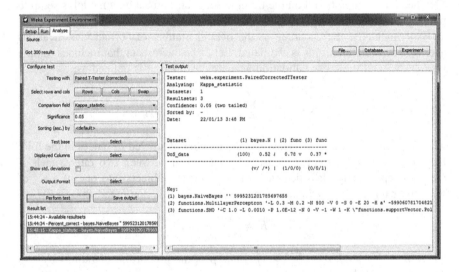

Fig. 8.20 DoS attack classification Kappa statistic of statistical based learning algorithms

among the rule based learning algorithms. J48 was the second choice within the rule based learning algorithms. In the Kappa statistics measure of Fig. 8.23, PART was again shown as the most superior. However, in terms of computational complexity measure, REPTree is the best choice as shown in Fig. 8.21. J48 was the second best choice comparing among these four rule based learning algorithms.

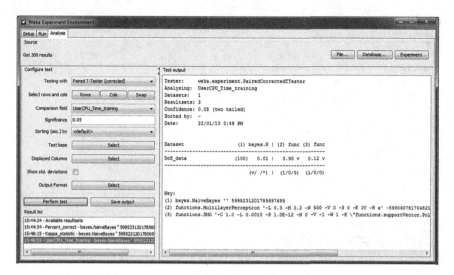

Fig. 8.21 DoS attack classification model building time of statistical based learning algorithms

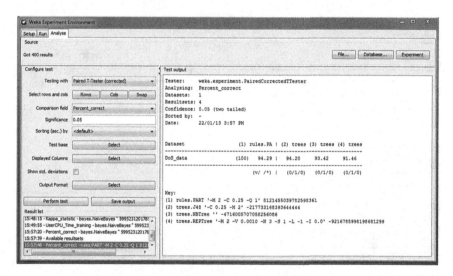

Fig. 8.22 DoS attack classification accuracies of rule based learning algorithms

From the above discussions, it can be concluded that PART algorithm is the best choice to prevent any type of DoS attack in the Smart Grid network. This was not only the fastest algorithm but also in terms of attack classification measures, it appears to be the most accurate. Basically, it uses separate-and-conquer method to build a model. It generates a partial C4.5 decision tree (which is implemented as J48 in Weka) in each iteration and makes the "best" leaf into a rule.

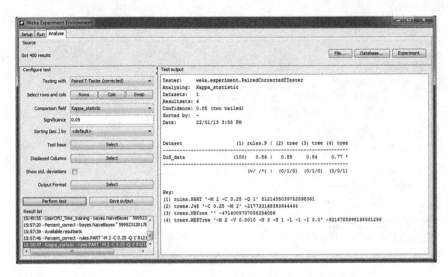

Fig. 8.23 DoS attack classification Kappa statistic of rule based learning algorithms

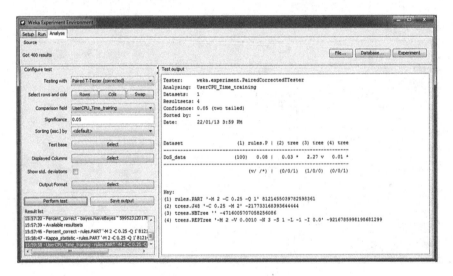

Fig. 8.24 DoS attack classification model building time of rule based learning algorithms

8.10 Discussions

Use of machine learning algorithms to foil DoS attack by means of a simple data-based approach for the Smart Grid network is comparatively a novel technique. In this chapter, the authors put forward a rule based learning approach, PART, for DoS attack classification problem. Unlike other rule based learners, it is easier to visualize the rules during the PART algorithm learning process. PART performance on the real life data were tested for preventing the Dos attack in the Smart Grid network and its performance was compared with that of other rule based learning algorithms. In addition, performance of PART was also tested against a group of statistical based learning algorithms. It is evident that PART outperforms other approaches on both prediction accuracy and the Kappa statistics measure. In terms of computational complexity measure, PART is not the best choice for preventing DoS attack, although its computational complexity is comparable to that of other rule-based algorithms. Further testing on other types of attack in the Smart Grid network is in progress in order to study the robustness of PART algorithm. The classification performance could be improved by adopting the best suited feature selection algorithm inside the PART implementation.

References

1. S. G. W. Group (2003) Challenge and opportunity: charting a new energy future: appendix A working group reports. Energy Future Coalition, Washington DC
2. Paro A, Fadigas E (2011) A methodology for biomass cogeneration plants overall energy efficiency calculation and measurement—a basis for generators real time efficiency data disclosure. In: Proceedings of power systems conference and exposition (PSCE), pp 1–7
3. Denholm P et al (2010) The role of energy storage with renewable electricity generation. National Renewable Energy Laboratory, Colorado
4. DeCarolis JF, Keith DW (2006) The economics of large-scale wind power in a carbon constrained world. Energy Policy 34:395–410
5. Archer CL, Jacobson MZ (2007) Supplying base load power and reducing transmission requirements by interconnecting wind farms. J Appl Meteorol Climatol 46:1701–1717
6. Freris L, Infield D (2008) Renewable energy in power systems. Wiley, New York
7. EDAI Department of Employment (2011) Queensland energy management plan, department of employment, economic development and innovation, Queensland government. http://rti.cabinet.qld.gov.au/documents/2011/may/qld%20energy%20management%20plan/Attachments/Qld%20Energy%20Mgt%20Plan.pdf. Accessed 13 Oct 2011
8. Delucchi M. A, Jacobson M. Z (2011) Providing all global energy with wind, water, and solar power, Part II: Reliability, system and transmission costs, and policies. Energy Policy 39:1170–1190
9. Grant W et al (2009) Change in the Air. Power Energ Mag IEEE 7:47–58
10. Zhong J et al (2010) Wind power forecasting and integration to power grids. In: Proceedings of 2010 international conference on green circuits and systems (ICGCS), pp 555–560
11. Sense of Security Pty Ltd (2011) Securing the Smart Grid. In: Proceedings of smart electricity world conference
12. Jamieson A (2011) Close the door! securing embedded systems. In: Proceedings of AusCERT information security conference
13. Smart Grid Security Myths vs. Reality (2012) White paper, SilverSpring Networks
14. Smart grid security critical success factors. http://www.cio.com.au/article/363005/smart_grid_security_critical_success_factors/R,Cited. 11 Feb 2013
15. McDowell M (2009) Understanding denial-of-service attacks. http://www.us-cert.gov/cas/tips/ST04-015.html. Accessed 10 Jan 2013
16. Ali ABMS (2012) What's at risk as we get smarter?. IEEE Smart Grid Newsletter, USA
17. Khorshed M T et al (2011) Monitoring insiders activities in cloud computing using rule based learning. In: Proceedings of IEEE trustcom-11, Changsha, China

18. Khorshed MT et al (2012) Classifying different DoS attacks in cloud computing using rule based learning, security and communication networks. Wiley, New York
19. Khorshed M T et al (2011) Trust issues that create threats for cyber attacks in cloud computing. In: Proceedings of IEEE ICPADS, Tainan, Taiwan
20. ecuritytube.net. (2012) Ddos attack with Rdos and T3c3i3. http://www.securitytube.net/video/471922. Accessed 12 Aug 2012
21. Batishchev AM (2012) LOIC. http://sourceforge.net/projects/loic/. Accessed 22 Aug 2012
22. G. Inc. (2012) NewEraCracker LOIC. https://github.com/NewEraCracker/LOIC/22. Accessed Aug 2012
23. BBC (2010) Anonymous wikileaks supporters explain web attacks. http://www.bbc.co.uk/news/technology-11971259. Accessed 23 Aug 2012
24. Khorshed MT et al (2012) A survey on gaps, threat remediation challenges and some thoughts for proactive attack detection in cloud computing, Future Generation Comput Syst Elsevier 28(6):833-851
25. Nanda R (2008) DDoS attack/PING flooding: explanation and solution. http://ramannanda.blogspot.com.au/2009/05/ddos-attackping-flooding-explanation.html. Accessed 23 Aug 2012
26. Grid G (2010) Tutorial: how to DoS attack (ping flooding). http://ghostgrid.blog.com/2010/12/16/ping-flooding/. Accessed 23 Aug 2012
27. Rouse M (2006) Ping of death. http://searchsecurity.techtarget.com/definition/ping-of-death. Accessed 23 Aug 2012
28. Kumar A et al (2012) Performance evaluation of centralized multicasting network over ICMP ping flood for DDoS, Performance Evaluation. Int J Comput Appl 37(10):1-6
29. Wilmes G, Kistler U (2007) Engage packet builder—scriptable libnet-based packet builder. http://www.engagesecurity.com/products/engagepacketbuilder/. Accessed 24 Aug 2012
30. John GH, Langley P (1995) Estimating continuous distributions in bayesian classifiers. In: Proceedings of 11th conference on uncertainty in artificial intelligence, San Mateo, pp 338–345
31. Michie D et al (1994) Machine learning, neural and statistical classification. Ellis Horwood series in artificial intelligence, Chichester, New York
32. Platt JC (1999) Fast training of support vector machines using sequential minimal optimization, Advances in Kernel Methods—Support Vector Learning, pp 185–208
33. Keerthi SS et al (2001) Improvements to platt's SMO algorithm for SVM classifier design. Neural Comput 13:637–649
34. Frank E, Witten IH (1998) Generating accurate rule sets without global optimization. In: Proceedings of 15th international conference on machine learning, pp 144–151
35. Quinlan JR (1993) C4. 5: programs for machine learning. Morgan Kaufmann, San Mateo
36. Kohavi R (1996) Scaling up the accuracy of naive-Bayes classifiers: a decision-tree hybrid. In: Proceedings of the 2nd international conference on knowledge discovery and data mining
37. Witten IH et al (2011) Data mining: practical machine learning tools and techniques: practical machine learning tools and techniques. Morgan Kaufmann, USA
38. Contextuall (2012) What is 10-Fold cross validation? https://contextuall.com/what-is-10-fold-cross-validation/. Accessed 12 Jan 2013
39. Cohen J (1960) A coefficient of agreement for nominal scales. Educ Psychol Measur 20:37–46
40. Marris E (2008) Upgrading the grid. Nature 454:570–573

Chapter 9
Smart Grid Communication and Networking Technologies: Recent Developments and Future Challenges

Faisal Tariq and Laurence S. Dooley

Abstract The smart grid is ostensibly the next generation power grid in which electrical energy distribution and management is efficiently performed by exploiting information communication technologies such as pervasive computing, in the control and decision-making processes. The smart grid is characterised by such functionality as being able to adapt to load and demand changes, intelligently manage bidirectional data flow and crucially enhance system reliability, robustness, security and sustainability. Communication networks play a crucial role in facilitating these features and are an integral component in any smart grid management system. In this chapter, the role of the communications network in smart grid operation is described together with its main functionalities. In particular, the challenges and opportunities for integrating existing and future wireless and mobile networks into the smart grid will be analysed, while the chapter concludes by identifying some future research directions for smart grid technologies.

9.1 Introduction

In the past few decades, electrical and electronic technologies have experienced massive advancement with the development of a myriad of commodities to make our daily lives more convenient and comfortable. These range for example, from smart phones and popular mobile consumer gadgetry to intelligent temperature and humidity control systems and everyday household appliances like cookers, dishwashers and fridges. Inevitably, this has led to electricity consumption commensurately increasing, with fossil fuels, coal, gas and nuclear power serving as the principal sources of electrical energy in most developed countries. These relatively

F. Tariq (✉) · L. S. Dooley
Department of Computing and Communication, The Open University,
Milton Keynes, UK
e-mail: ftariq@ieee.org

A. B. M. S. Ali (ed.), *Smart Grids*, Green Energy and Technology,
DOI: 10.1007/978-1-4471-5210-1_9, © Springer-Verlag London 2013

cheaper sources are often linked to a corresponding detrimental effect upon the environment and global climate. Society has also become increasingly aware of the consequences of climate change, to the extent that new cleaner renewable energy sources are being sought as expedient solutions for the future supply of electrical energy [1, 2], with wind, wave and solar power being the most representative examples of these newer sources of energy.

A corollary of this diversification of energy sources is that electric distribution systems will be transformed from the current situation of having a relatively small number of very large generation plant–based systems, to a much more heterogeneous-based framework combining the traditional large-scale grid with emerging micro-grids of solar, wave and wind power, distributed over geographical areas. The energy distribution and network information flow will thus inexorably shift from the present unidirectional paradigm to a more flexible bidirectional model with the management requirements being inevitably more complex within existing infrastructures. As existing strategies also become increasingly outmoded for managing the emergent heterogeneous electricity grids, the quest for new ways to handle future growth in demand and response will be paramount, allied with the obligation for such new systems to be reliable, robust, sustainable, safe and secure.

The smart grid [3, 4] is the next generation power distribution network which has the overarching aim of transforming the current centralised energy producer control grid into a more distributed user interactive power system. The USA Department of Energy envisages the smart grid as applying technologies, tools and techniques to bring intelligence to the power grid [5]. Amongst its key features will be that it is reliable, affordable and secure, as well as having the capability of incorporating heterogeneous energy supplies including renewable energy sources while concomitantly striving to minimise its carbon footprint.

To fulfil the vision and realise the potential of the smart grid, a sophisticated communications network will inevitably play a pivotal role in any smart grid management system [6]. It is for this reason that an investigation into the requirements and challenges in both designing and implementing an effective smart grid communication and network system is essential.

The remainder of this chapter is organised as follows: The communication network infrastructure for the so-called *legacy grid* will firstly be described along with a review of some of the limitations in its applicability to future power grids. A synopsis of the communication and networking requirement for future smart grids will then be presented before networking, and topological-related issues are discussed along with key security implications and standardisation initiatives in the domain. Finally, the chapter concludes by proposing some future research directions for smart grid technologies.

9.2 Communication and Networking in the Legacy Power Grid

The legacy power grid generally consists of a few, very large power generating stations which supply power to an entire region or even country, via a huge network of transformers and cables, with the distribution inherently being unidirectional in nature. The power grid has a rigid hierarchical structure, with the power generation plants being at the top of the system, while the load at the customer premises is at the bottom [7].

Energy demand is normally modelled based upon the peak time requirement. The generation capacity therefore must be greater than this peak demand since it is simply not financially feasible to store electric energy in large quantities [8]. As the average energy demand is typically much lower than the peak, the capacity of the electric generation system remains seriously underutilised for a considerable proportion of the time. Under the existing centralised distribution and static power grid management system, there is little flexibility to be able to distribute some of the excess peak load in these off-peak periods. To achieve this more efficient paradigm, greater interactivity and information sharing from the end-users is required. However, existing communication and management systems were never designed with this laudable aim in mind.

The three principal functions of the communication and networking system in a legacy power grid are (1) data acquisition from sensors located at different distribution points along the grid; (2) transmitting command and control signals to sensors and actuators; and (3) fault detection in the power generation system and distribution networks. The most widely adopted system for both monitoring and controlling these industrial communication processes is the software package known as *Supervisory Control And Data Acquisition* (SCADA) system [9].

9.2.1 SCADA System

SCADA utilises *programmable logic controllers* (PLC) as the interface for control management and is able to handle anywhere in the range from a few hundred to one million or more I/O (*input/output*) devices in the field [10].The software has two layers. The first is the *client layer* which is used for human operator interfacing, and the second is the *data server layer* which manages the control processes of the system. SCADA operates in a multitasking fashion based on a real-time database system located in different servers.

SCADA is a large-scale operation which requires constant monitoring, supervision and safeguards, since the consequences of failure can be devastating on the entire electric supply system. Studies on major blackouts in certain Western countries over the past decade have revealed that the system responses were either too slow or, in other cases, unable to control or repair the damage caused [11].

Many operators employ open-source versions of the SCADA software for financial expediency, though these versions are acknowledged to have loopholes [12], which pose a potentially serious security risk.

9.2.2 Communication and Networking in the Smart Grid

As the emerging smart grid will deploy a vast number of sensors and actuators at the user equipment level to both collect usage data and to deliver command and control information, the current legacy communication system is incapable of supervising such a demanding scenario. Consequently, new communication system architectures and tools need to be developed [13].

Future smart grids will be both intelligent and adaptive to the changes in the scenario and so will require instantaneous acquisition and transmission of data from different entities in the system. This means near real-time secure communications is a vital design requirement to convey command and control information between different entities in the smart grid. The constituent components of a smart grid can be broadly classified into three categories: (1) generation and supply components, (2) transmission and distribution components, and (3) consumer equipment, with the underlying requirements for the communication technologies pragmatically being different for each entity [14, 15]. For example, home appliances may only occasionally require data transfer when operating, while conversely, a nuclear power generation plant requires continual monitoring of its data communications.

The most important requirement of a communication system is that it should be reliable and available everywhere at any time. In order to ensure reliable transfer of time critical information, all the entities must be contactable regardless of the situation and distances involved. The system must also be capable of automatically managing redundancy and be adaptive to changes in the network topology and surrounding environment. The information to be transferred by the communication system includes consumer usage, billing, generation and load, which all contain critical and sensitive data which must be protected. This means the system has to be sufficiently robust and secure to maintain the necessary levels of privacy. Today's typical household has only one or two devices connected to the Internet, but with the smart grid, there will many more connected, with a commensurate rise in the amount of data to be transferred over the network. The communications system has therefore be able to support commensurately high bandwidths.

Another desirable feature for the communications system is that it includes a self-organising capability which will ensure healing from potential faults or network problems with minimum manual and central intervention. A benefit of having a latent self-organising feature is that it will also significantly lower the maintenance costs for the system [16].

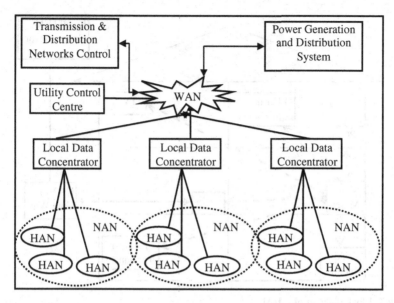

Fig. 9.1 Schematic diagram of the communication networks in the emerging smart grid

9.2.3 Network Topologies

From a networking topological viewpoint, smart grid communication entities can be divided into three types of network architecture, namely the *Home Area Network* (HAN), the *Neighbouring Area Network* (NAN) and the *Wide Area Network* (WAN). Figure 9.1 shows a schematic diagram of the communication network for the smart grid, with each block now being individually described.

9.2.3.1 Home Area Network

This is the basic core element of smart grid networking systems [17]. Figure 9.2 shows a block diagram of a HAN, with energy-consuming home appliances and energy meters being the main constituent nodes. Ideally, all home appliances will be connected to the smart meter which will transfer the measurement data firstly to the home hub and then onto the central management entity.

There are several available technologies which facilitate communications between appliances, provided sensors have either been installed or are embedded within the appliances for inter-device communications. Zigbee [18] is the most prominent technology in this respect, while for inter-device communications, *machine to machine* (M2M) technology [19, 20] can be exploited, with M2M currently being standardised in the LTE release 10 and beyond [21].

An entity is required to transfer the data collected by the smart meter to the local data concentrator. Both *Wireless Local Area Network* (WLAN) and the

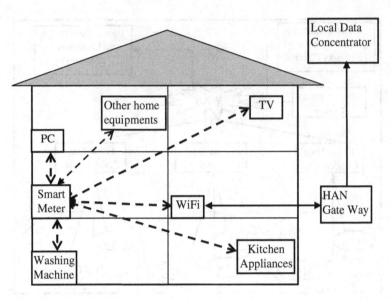

Fig. 9.2 Block diagram of a HAN

emerging femtocell technology [22] are feasible options as home communication hubs, with the former using IP-based Ethernet and the latter a combination of wired backhaul and cellular network for data backhauling to the distributor.

The HAN is typically managed by a *home energy management system* (HEMS) which allows the energy management process to be more interactive and dynamic [23]. It will help to optimise energy usage to ensure a minimum bill for the consumer. For example, some tasks require higher amounts of energy but do not need to be immediate executed, so they can be automatically scheduled at a time when the energy tariff is lower, that is, during the night. Another example is the emerging *plug-in hybrid electric vehicle* (PHEV), where the car can be recharged during the night when the electricity spot price falls to a certain value which is able to be preset by the user.

9.2.3.2 Neighbouring Area Network

These consist of several HANs which are in-turn connected a *local data concentrator* (LDC). The main role of the NAN is to reliably collect energy consumption information from the HANs and store it in the LDC. The LDC has responsibility for accumulating data from various meter readings at different times of the day. It also plays a vital role in managing demand response because it enables the central entity to exploit the data by interpreting usage patterns, so that appropriate billing strategies can be developed to distribute demand across a defined time period.

The coverage of the NAN may include a few buildings in an urban area with each building consisting of several flats or apartments. In rural areas, the NAN would include several geographically co-located houses. The actual area covered by NAN varies depending upon the scenario. Analysis in [24] suggests that the NAN coverage should include buildings which are up to a maximum of 500 and 700 m away from the LDC in urban and rural areas, respectively. In urban areas, WLAN can be employed for communications amongst the entities of the NAN, with femtocell networks an attractive alternative, while in rural areas, cellular systems can be exploited for communications between the NAN entities.

9.2.3.3 Wide Area Networking

WAN consists of several HAN and NAN and also include other entities like LDC and the central controller. WAN is spread over an extensive geographical area and maintains communications between all the other entities such as the generation and distribution units. WAN is effectively the communication backbone for the smart grid communication system, with Ethernet-based IP networks, cellular networks, *Worldwide interoperability for Microwave Access* (WiMAX), and long-range microwave transmissions all possibly forming part of the WAN in a future smart grid framework.

Since different types of wireless and wired networks may be part of the WAN, interoperability is extremely important so data from the sensors and actuators can flow seamlessly throughout the networks. Furthermore, the route needs to be optimised in the presence of multiple transmission options. In most cases, the smart grid will be sharing the network with other applications and so may encounter a potential delay and/or cause network congestion. Smart grid data also may include time critical information, so there has to be either a dedicated network or alternatively some prioritised channel mechanism to ensure timely delivery of all time critical data.

9.2.4 Communication Technologies for the Smart Grid

A variety of diverse communication technologies are available to perform different tasks within the smart grid, ranging from usage data collection and transmission, through to monitoring generation and distribution. Some of the most promising technologies will now be reviewed.

9.2.4.1 ZigBee

This is a set of specifications designed to connect a variety of devices under single network control. It is aimed at serving low-cost, low-powered wireless sensor networks, with the latest version of ZigBee being capable of accommodating more than 64,000 devices in a single network. It is widely adopted for various applications including building automation, remote monitoring and sensing, and meter reading [25].

A separate version for energy applications known as *ZigBee Smart Energy* has been developed recently. Its numerous features include dynamic pricing enhancement, prepayment option, tunnelling of other protocols and over-the-air updates [26]. ZigBee operates on the unlicensed 868 MHZ, 915 MHz and 2.4 GHz bands and can typically support from 20 to 250 Kbps data rate within the coverage of up to 100 m. It uses *direct sequence spread spectrum* (DSSS) modulation and supports various network topologies including tree, star and mesh. The security provision for ZigBee is maintained using 128 bit *advanced encryption system* (AES). Despite its many advantages, the technology has some shortcomings that have hampered its widespread adoption. Principally amongst this is the limited battery lifetime are due to its small size, ZigBee may not be suitable for many smart energy applications which require continuous monitoring and a heavy processing capacity. Also if the data rate requirement is too high, then ZigBee devices may not be the best choice.

9.2.4.2 WLAN

In contrast, IEEE 802.11 is a set of standards defined for WLAN to support both point-to-point and point-to-multipoint communications. The earliest version of the standard was released in 1997 and subsequent amendments have been proposed, so that currently a number of WLAN versions are available [27]. WLAN can be a particularly advantageous solution for smart grid data communications for a number of reasons. Firstly, the device is widely accepted and used around the world for Internet access with many households already having one installed. Secondly, WLAN is relatively cheap, so even if a new installation is required, it will not be too costly; and thirdly, the WLAN card is a plug-and-play device and so is straightforward to install.

Earlier versions of this standard used both DSSS and *frequency-hopped spread spectrum* (FHSS) modulation, though latter variants have employed *Orthogonal Frequency-Division Multiplexing* (OFDM) because of its intrinsic spectral efficiency and provision for supporting data rates of several Mbps. IEEE 802.ad which is the most recent standardising technology also uses OFDM together with *Multiple Input Multiple Output* (MIMO) and operates on a 60-GHz band, with the promise of a maximum data rate of up to 7 Gbps [28].

WLAN typically operate on either the 2.4- or 3.5-GHz bands which are shared with some other standards, so it can be prone to severe interference in densely deployed areas. If there are other neighbouring devices operating on the same band, then its performance can degrade significantly, with reliability therefore being undermined. This is undoubtedly the leading challenge facing WLAN-based communications for adoption in smart grid systems [29].

9.2.4.3 WiMAX

This is part of the IEEE802.16 series of standards and affords both fixed and mobile broadband services. WiMAX normally operates on either a 2.5- or 3.5-GHz channel, though the standard supports between 2 and 11 GHz, and 10 to 66 GHz transmissions, with a scalable bandwidth in the range from 1.25 to 20 MHz at a peak data rate of 70 Mbps and a coverage range of up to 50 km. However, data rates are much lower at the cell edge due to severe interference. Some recent WiMAX standards offer peak data rates of around 1 Gbps for stationary users [30].

WiMAX is a promising technology for smart grid data communications. Fixed WiMAX supports high data rates and can be used for data backhauling if wired networks are either unavailable or congested due to high traffic loads. Mobile WiMAX can be used for collecting meter reading data from the HAN and also for conveying real-time pricing information from the central controller to the HEMS. Another important aspect of the WiMAX network is outage detection in various entities in the system. It can also be used for optimally bringing the deviated voltage level up to a predefined threshold [31].

While WiMAX is attractive in terms of the functionalities it offers the smart grid, it concomitantly has a number of drawbacks. WiMAX networks are predominantly deployed to provide broadband Internet services to households and mobile users. This implies the network topology will not necessarily be optimal or even congruent for smart grid applications. Conversely, while a customised network for the smart grid system may be the most desirable solution to the problems highlighted above, it will also incur prohibitively high deployment costs, and the "last mile" to the consumer may experience outages due to severe interference.

9.2.4.4 Cellular Networks and Femtocells

The ubiquitous nature of cellular networks all around the world makes them an especially attractive option for smart grid communications. In particular, the emerging *3rd generation partnership project–long term evolution* (3GPP-LTE) standard is highly relevant to smart grid applications for a number of reasons. It offers high data rate services in the range of several Mbps for both stationary and mobile users [32, 33]. The recent development of indoor base stations popularly known as *femtocell access points* (FAP), which backhauls data via a wired network can now offer very high data rates. The FAP is a low-cost plug-and-play device which aims to extend cellular radio coverage within indoor environments where conventional mobile coverage typically deteriorates because of the poor quality signal caused by high wall penetration losses.

In addition to delivering all the functionality described for the WLAN, FAPs offer a number of other propitious advantages. For instance, unlike WLAN, FAPs operate on a licensed band so the *Quality of Service* (QoS) can be more consistently ensured by intelligently allocating channels. This means time critical data can be safely transmitted via femtocell networks. Another advantage is that the

208 F. Tariq and L. S. Dooley

same protocol can be used throughout the route, since the cellular network alone serves as WAN. For the interested reader, further details on FAP and cognitive femtocell technologies can be found in [34].

The superior QoS is not achieved without some cost, since with spectrum very expensive, the data transfer overheads will inescapably be higher than WLAN. So, in conclusion, cellular networks may be neither a cost-effective nor viable option for smart grid communications. It is also anticipated the smart grid will generate huge amounts of data to be transferred, so overloading an already saturated network. Thus, the capacity of the cellular network will have to improve in order to accommodate the demands of the smart grid.

For completeness, in addition to the aforementioned technologies, there are other candidate technologies worth mentioning for possible smart grid communications including Bluetooth, digital microwave and IEEE 802.20-based *mobile broadband wireless access*. These all provide to some degree, certain smart grid functionality though they are beyond the scope of this chapter.

9.2.5 Standardisation Activities

To ensure interoperability and seamless data flow between the devices of the various communication technologies reviewed in the previous sections, it essential a formal standardisation framework is established to which all interested parties adhere. There are many well-established standardisation development organisations such as the *Institute of Electrical and Electronics Engineers* (IEEE), *International Standard Organisation* (ISO) and *International Telecommunication Union* (*ITU*) which are collaborating to regulate smart grid communication technologies [35].

There are similar bodies working towards a standardisation for the smart metering technology. In Europe, for instance, *European Committee for Standardisation* (CEN), *European Committee for Electrotechnical Standardisation* (CENELEC) and *European Telecommunications Standardisation Institute* (ETSI) are all striving to create a new standard for enabling interoperability of various utility meters. ETSI is also working towards developing and maintaining a high level, end-to-end architecture for M2M communications, while the European Commission has announced a mandate M/441 for the creation of an open architecture for utility metering and their interoperability.

In the United States, IEEE P2030 aims to integrate energy technology and information and communications technology for enabling seamless communications amongst the different entities of the smart grid, while in the world's most populace country, the *State Grid Corporation of China* (SGCC) is overseeing the standardisation process [36]. Table 9.1 provides a summary of the current standards relating to smart grid communications and networking.

Table 9.1 Some smart grid–related standards

Standard code	Scope
ISO/IEC 14543-3 KNX, Open HAN, IEEE 802.15.4 ZigBee, ISO/IEC 18012, 6LoWPAN	HAN, building automation and smart metering
IPv4, IEC 62056, PLC G3, DNP3	Core communication networks including HAN, NAN and WAN

9.2.6 Research Challenges

Communication technologies and networks will clearly play a pivotal role in realising the full potential of the smart grid, though it is readily apparent from the above discussion that existing network technologies, which have ostensibly been developed for other applications, have neither the flexibly nor full functionality to be directly applied to the smart grid domain. New innovative communication and networking techniques and tools will need to be developed to accommodate the requirements of the smart grid, and this section will briefly investigate some of the key research challenges to be addressed if the smart grid vision is to be fulfilled.

9.2.6.1 Smart Home Networks

Research into home networking has tended to concentrate on enabling multimedia communications, such as audio/video streaming [37]. Existing home networks are normally capable of handling only a limited number of nodes, while the smart grid may introduce up to 30 nodes, with a commensurate increase in the complexity of both network connectivity and inter-device communications. Moreover, to analyse the energy consumption patterns of individual consumers and their equipment, meter readings will be taken at a greater frequency compared to the legacy grid, so there will be an equivalent rise in the data rates. To accommodate this large data rate demand, higher capacity home networking equipment will be needed.

Since most smart grid devices will be wireless, when their deployment density is high, mutual interference can become problematic. Strategies for either alleviating or minimising interference will therefore be an especially important research aim. To compound this issue, in densely populated urban areas, the coverage of neighbouring HAN will often overlap so generating interference between the HANs. While intra-HAN interference can be managed by negotiation, inter-HAN interference is a far more intractable difficult and will entail the design of new effective inter-HAN interference mitigation schemes.

9.2.6.2 Seamless Interoperability

Unlike the legacy grid, the smart grid includes an assortment of energy sources which are distributed over a wide geographical area. Traditional unidirectional flow models are no longer applicable, and the smart grid will most probably include a variety of entities belonging to different manufacturers. A common communication interface thus needs to be established which is independent of the manufacturers and works on all communication media.

A further challenge is to enable communication and data flow between a variety of diverse networks. For example, sensor networks communicate differently to cellular networks, so the internetwork interface has to be designed to ensure that the data can be seamlessly transferred from the consumer to the central controller over multiple heterogeneous networks.

9.2.6.3 Transmission Network and Route Optimisation

The smart grid will most probably share existing wired and wireless communication infrastructure for data transfer. The legacy grid network design is not optimal, and as a consequence, efficiently accessing the network may become compromised. There will also be a number of different network options for data transmission within the smart grid. While having such multiple options enhances the reliability of the communications, it also makes routing more convoluted. Appropriate route optimisation schemes will need to be developed to ensure maximum security at minimal cost. Cognitive routing algorithms are one possible alternative by monitoring and analysing key network parameters such as latency, hop count, length and traffic load to help solve what is in essence, a multi-objective function optimisation.

9.2.6.4 Security and Privacy

Security and privacy are a particular anxiety in smart grid communications [38] given that data are collected much more frequently than in the legacy grid to provide better pricing and load distribution. With the system being automated and flexible for remote control and dynamic management, it will become inevitably more vulnerable to cyber-attacks. Robust security measures are therefore essential to avoid the risk of large-scale damage, malicious data eavesdropping and wormholes. The blocking of unauthorised data access is a further worry, since public networks are to be used for data transmission, so the smart grid will not have full control over the network integrity. This security risk is exacerbated when the network is shared between multiple applications.

One ethical issue given the huge volumes of data being transferred is the privacy of the consumer, which requires protective measures to be taken to control the access to this data. Authentication mechanisms will have to be integrated into the system at various access points to ensure both security and privacy. Possible strategies to achieve this include special end-to-end encryption schemes whenever data are routed across public networks. Another alternative is to make all energy consumption data anonymous, so unauthorised intruders or eavesdroppers at any network access point are not capable of extracting either the customer's identity or their usage patterns.

9.2.6.5 Perspective of Developing Countries

Existing communications infrastructure in many developing countries, such as Bangladesh, India and Nigeria, is simply not suitable for data transfer on such a large scale as mandated by the smart grid with many lacking the requisite fibre optic–based wired backbones that are taken for granted in the West. Wired infrastructure is not a viable option then for smart grid communications while corresponding wireless networks are predominantly based on *second generation* (2G) technology, which is appropriate for only voice and low data rate applications. A few countries are installing *third generation* (3G) networks, though their coverage penetration is largely limited to urban areas. Data transmission over both 2G and 3G networks will also be very expensive, so the challenges to design and roll-out an appropriate cost-effective communications infrastructure to support smart grid applications [39] in these countries are immense.

9.3 Conclusion

The communications network will be an integral component in a future smart grid system, with its performance being dependent on data transfer capacity. The availability of near real-time data will allow the system to make better decisions with potential flow on benefits to the consumer. To ensure data availability whenever and wherever required, robust and seamless communications are required and the capability of the existing communications infrastructure needs to be upgraded to accommodate this massive data growth. The chapter has provided an overview of recent smart grid communications technology developments and their associated challenges for ensuring successful deployment with respect to the key issues of reliability, versatility, privacy and security. Some of the current standardisation initiatives were also examined, and the chapter concluded with a discussion on the state-of-the-art research challenges and opportunities for integrating existing and future wireless networks with smart grid technologies.

References

1. Bilgen S, Kaygusuz K, Ahmet S (2004) Renewable energy for a clean and sustainable future. Energy Sources 26(12):1119–1129
2. Turner JA (1999) A realizable renewable energy future. Science 285(5428):687–689
3. Momoh J (2012) Smart grid: fundamentals of design and analysis. Wiley, NY
4. Gharavi H, Ghafurian R (2011) Smart grid: the electric energy system of the future. Proc IEEE. 99(6):917–921
5. US Department of Energy (2009) Smart grid: an introduction. http://energy.gov/oe/downloads/smart-grid-introduction. Cited 04 Nov, 2012
6. Bouhafs F, Mackay M, Merabti M (2012) Links to the future: communication requirements and challenges in the smart grid. IEEE Power Energ Mag 10(1):24–32
7. Farhangi H (2010) The path of the smart grid. IEEE Power Energ Mag 8(1):18–28
8. Ibrahim H, Ilinca A, Perron J (2008) Energy storage systems—characteristics and comparisons. Renew Sustain Energy Rev 12(5):1221–1250
9. Daneels A, Salter W (1999) What is SCADA. international conference on accelerator and large experimental physics control systems, pp 339–343
10. Fernandez JD, Fernandez AE (2005) SCADA systems: vulnerabilities and remediation. J Comput Small Coll 20(4):160–168
11. Amin MS, Wollenberg BF (2005) Toward a smart grid: power delivery for the 21st century. IEEE Power Energ Mag 3(5):34–41
12. Igure VM, Laughter SA, Williams RD (2006) Security issues in SCADA networks. Comput Secur 25(7):498–506
13. Wu FF, Moslehi K, Bose A (2005) Power system control centers: past, present, and future. Proc IEEE 93(11):1890–1908
14. Hu R, Yi Q, Wang J (2012) Recent advances in wireless technologies for smart grid [Guest Editorial]. IEEE Wirel Commun 19(3):12–13
15. Gungor VC, Sahin D, Kocak T, Ergut S, Buccella C, Cecati C, Hancke GP (2013) A survey on smart grid potential applications and communication requirements. IEEE Trans Ind Inf 9(1):28–42
16. Sauter T, Lobashov M (2011) End-to-end communication architecture for smart grids. IEEE Trans Ind Electron 58(4):1218–1228
17. Zhou C, Hassanein HS, Qiu R, Samarati P (2012) Communications and networking for smart grid: technology and practice. International journal digital multimedia broadcasting, 2011
18. Parikh PP, Kanabar MG, Sidhu TS (2010) Opportunities and challenges of wireless communication technologies for smart grid applications. IEEE Power and Energy Society General Meeting, pp 1–7
19. Shrestha GM, Jasperneite J (2012) Performance evaluation of cellular communication systems for M2M communication in smart grid applications. Computer Networks, pp 352–359
20. Salam SA, Mahmud, SA, Khan GM, Al-Raweshidy HS (2012) M2M communication in smart grids: implementation scenarios and performance analysis. In: IEEE wireless communications and networking conference workshops (WCNCW), pp 142–147
21. Lu G, Seed D, Starsinic M, Wang C, Russell P (2012) Enabling smart grid with ETSI M2M standards. In: IEEE wireless communications and networking conference workshops (WCNCW), pp 148–153
22. Tariq F, Dooley LS, Poulton AS (2011) Virtual clustering for resource management in cognitive femtocell networks. 3rd international congress on ultra modern telecommunications and control systems, pp 1–7
23. Inoue M, Higuma T, Ito Y, Kushiro N, Kubota H (2003) Network architecture for home energy management system. IEEE Trans Consum Electron 49(3):606–613

24. Lopez G, Moura PS, Custodio V and Moreno JI (2012) Modeling the neighborhood area networks of the smart grid. IEEE international conference on communications (ICC), pp 3357–3361
25. Wheeler A (2007) Commercial applications of wireless sensor networks using ZigBee. IEEE Commun Mag 45(4):70–77
26. http://www.zigbee.org/Specifications.aspx. Cited 10 Feb 2013
27. Kuran MS, Tugcu T (2007) A survey on emerging broadband wireless access technologies. Comput Netw 51(11):3013–3046
28. Vaughan-Nichols SJ (2010) Gigabit Wi-Fi is on its way. Computer 43(11):11–14
29. IEEE 802 Standard (2011) IEEE Standard for Information technology—Local and metropolitan area networks—Specific requirements—Part 11, pp 1–433
30. Sood VK, Fischer D, Eklund JM, Brown T (2009) Developing a communication infrastructure for the smart grid. In: IEEE electrical power and energy conference (EPEC), pp 1–7
31. Mao R, Li H (2012) An efficient multiple access scheme for voltage control in smart grid using WiMAX. IEEE international conference on communications (ICC), pp 367–3371
32. Du J, Manli Q (2012) Research and application on LTE technology in smart grids. International conference on communications and networking in China (CHINACOM), pp 76–80
33. Cheng P, Wang L, Zhen B, Wang S (2011) Feasibility study of applying LTE to Smart Grid. IEEE international workshop on smart grid modeling and simulation (SGMS), pp 108–113
34. Tariq F, Dooley LS (2012) Cognitive femtocell networks. In: Grace D, Zhang H (eds) Cognitive communications: distributed artificial intelligence (DAI), regulatory policy and economics, implementation. Wiley, NY, pp 359–394
35. Kayastha N, Niyato D, Hossain E, Han Z (2013) Smart grid sensor data collection, communication, and networking: a tutorial. Wireless communications and mobile computing, pp 1530–8677. DOI:10.1002/wcm.2258
36. Fan Z, Kulkarni P, Gormus S, Efthymiou C, Kalogridis G, Sooriyabandara M, Zhu Z, Lambotharan S, Chin W (2013) Smart grid communications: overview of research challenges, solutions, and standardization activities. IEEE Commun Surv Tutorials 15(1):21–38
37. Alam MR, Reaz MBI, Ali MAM (2012) A review of smart homes—past, present, and future. IEEE Trans Syst Man Cybern Part C Appl Rev 42(6):1190–1203
38. Wang X, Yi P (2011) Security framework for wireless communications in smart distribution grid. IEEE Trans Smart Grid 2(4):809–818
39. Jamal T, Ongsakul W, Lipu MS, Howlader MM, Aziz-Un-Nur IS (2012) Development of smart grid in Bangladesh: challenges and opportunities. In power and energy systems. ACTA Press, Calgary, Canada

Chapter 10
Economy of Smart Grid

Gang Liu, M. G. Rasul, M. T. O. Amanullah and M. M. K. Khan

Abstract Smart grid is generally characterized by high installation cost and low operating costs. Thus, the basic economic analysis is the one comparing an initial known investment with estimated future operating costs. Most smart gird requires an auxiliary energy source so that the system includes both renewable and conventional equipment, and the annual loads are met by a combination of the sources. In essence, renewable-based smart grid is bought today to reduce tomorrow's electricity bill. The costs of smart grid include all items of hardware and labor that are involved in installing the equipment plus the operating costs. Factors which may need to be taken into account include capital cost, replacement cost, and operating and maintenance (O&M) costs, insurance, fuel, and other operating expenses. The objective of the economic analysis can be viewed as the determination of the least cost method of meeting the energy need, considering both renewable and non-renewable alternative. In this chapter, several ways of doing economic evaluation, with emphasis on the life cycle savings method are noted. This method takes into account the value of money and allows detailed consideration of the complete range of costs. In this chapter, the costs of smart grid, economic indicators of smart grid, design variables of smart grid, and a case study in Central Queensland are presented.

G. Liu (✉)
School of Energy Science and Engineering, Central South University, Changsha, 410004 Hunan, China
e-mail: gliu001@hotmail.com

G. Liu · M. G. Rasul · M. T. O. Amanullah · M. M. K. Khan
Power and Energy Research Group, Faculty of Sciences, Engineering and Health, Central Queensland University, Queensland 4702, Australia

A. B. M. S. Ali (ed.), *Smart Grids*, Green Energy and Technology,
DOI: 10.1007/978-1-4471-5210-1_10, © Springer-Verlag London 2013

10.1 Costs of Smart Grid

Initial capital costs in buying and installing solar energy equipment are important factors in smart grid. They consist of the delivered price of components such as solar PV panels, meters, and inverters. Costs of installing these components must also be considered, as these can match or exceed the purchase fees.

Installation costs of solar equipment can be shown as the sum of two terms, one proportional to the size of components and the other independent of size [1]:

$$C_S = \sum_{i=1}^{N} C_{A,i} S_{C,i} + C_E \tag{10.1}$$

where
C_S Total cost smart grid ($)
C_A Cost of individual components ($)
S_C Sizes of component (-)
C_E Cost of installed smart grid ($)
i The number of components (-)

The size-dependent costs C_A include such items as the purchase and installation of the components and a portion of storage costs. The size-independent costs C_E include such items as controllers and erection equipment or bringing construction to the site, which do not depend on the size of components.

The **replacement cost** is the cost of replacing a component at the end of its lifetime. This is different from the initial capital cost for several reasons:

- Not all of the components may require replacement at the end of its life. For example, the wind turbine nacelle may need replacement but the tower may not.
- The initial capital cost may be reduced or eliminated by a donor organization, but the replacement cost may not.
- The fixed costs (e.g., travel cost) of a visit to the site are accounted. At initial construction, these costs are shared by all components, but at replacement time they may not.
- A reduction over time in the purchase cost of a particular technology is accounted.

Operating costs are associated with smart grid. The continuing costs include cost of auxiliary energy, energy costs for operation of all devices, extra real estate taxes imposed on the basis of additional assessed value of building or facility, interest charges on any funds borrowed to purchase the components, and others. Income tax implications might be included in the purchase of equipments, components, device, or other facilities.

For smart grid in this book, the electricity bill caused by purchasing electricity from the grid is also considered as the operating cost.

The concept of **smart savings**, as a useful economic indicator, is the difference between the cost of a conventional system and a smart grid.

$$S_{SG} = C_{CG} - C_{SG} \tag{10.2}$$

where

S_{SG} Smart savings ($)
C_{CG} Cost of conventional grid ($)
C_{SG} Cost of smart grid ($)

10.2 Economic Indicators of Smart Grid

The net present cost (NPC) of smart grid is the sum of the present value of all costs over the period of interest. If a number of system options are being considered, then the option with the lowest NPC is the most favorable financial option.

It includes residual values as negative costs. The total NPC of a project is a summation of all cost components including:

- Capital investment
- Non-fuel operation and maintenance (O&M) costs
- Replacement costs
- Energy costs (fuel cost plus any associated costs)
- Any other costs, such as legal fees, etc.

The present value is the value of a future transaction discounted to some base date. It reflects a time value for money. The present day equivalent of a future cost, that is, the present value, can be thought of as the amount of money that would be invested today, at an interest rate equal to the discount rate, in order to have the money available to meet the future cost at the time when it is predicted to occur [2].

$$\text{Present Value} = \frac{F_n}{(1+r)^n} \tag{10.3}$$

where

F_n Cash flow ($)
r The discount rate (%)
$1 + r$ The single payment presents worth factor
n Number of years (-)

Lifecycle cost (LCC) refers to the total cost of ownership over the life of a smart grid. It is the sum of all the costs associated with a smart grid over its lifetime or over a selected period of analysis, in today's dollars, and takes into account the time value of money. Typical areas of expenditure which are included in calculating the whole-life cost include planning, design, construction and acquisition, operations, maintenance, renewal and rehabilitation, depreciation and cost of finance, and replacement or disposal [3].

Annualized lifecycle cost (ALCC) is defined as the average yearly outflow of money (cash flow). The actual flow varies with year, but the sum over the period of an economic analysis can be converted to a series of equal payments in today's money that are equivalent to the varying series [3].

The **cost of energy (COE)** equation is one analytical tool that can be used to compare alternative technologies when different scales of operation, investment, or operating time periods exist. For example, the COE could be used to compare the COE generated by a PV power plant with that of a fossil fuel–generating unit or another renewable technology.

The calculation for the COE is the net present value of total life cycle costs of the system divided by the quantity of energy produced over the system's life [3].

$$\text{COE} = \frac{\text{Total life cycle cost}}{\text{Total life cycle energy consumption}} \qquad (10.4)$$

Return of investment (ROI), in finance, is the ratio of money gained or lost (whether realized or unrealized) on an investment relative to the amount of money invested. The amount of money gained or lost may be referred to as interest, profit/loss, gain/loss, or net income/loss. The money invested may be referred to as the asset, capital, principal, or the cost basis of the investment. ROI is usually expressed as an arithmetic return by [3]:

$$\text{ROI} = \frac{V_f - V_i}{V_i} \qquad (10.5)$$

where
V_f Gain from investment ($)
V_i Cost from investment ($)

Payback time (PBT) is a measure of how quickly the cash flow generated by the system covers the initial investment. Investors obviously prefer a shorter to a longer PBT. It is defined in many ways [3]:

- The time needed for the cumulative fuel savings to equal the total initial investment, that is, how long it takes to get an investment back by savings in fuel.
- The time needed for the yearly cash flow to become positive.
- The time needed for the cumulative savings to reach zero.
- The time needed for the cumulative savings to equal the down payment on the smart gird.
- The time needed for the cumulative smart savings to equal the remaining debt principal on the smart gird.

The **cash flow** of the smart grid shows a graph of the cash flow of the smart grid (Fig. 10.1). Each bar in the graph represents either a total inflow or total outflow of cash for a single year. The first bar, for year zero, shows the capital cost of the system, which also appears in the optimization results. A negative value represents an outflow or expenditure for fuel, equipment replacements, or O&M. A positive value represents an inflow, which may be income from electricity sales or the salvage value of equipment at the end of the project lifetime [5].

The cash flows can be displayed as either nominal or discounted values. A nominal cash flow is the actual income minus cost that the grid works in a particular year. A discounted cash flow is the nominal cash flow discounted to year zero. The discounted cash flow is calculated by multiplying the nominal cash flow by the discount factor.

There are three options for displaying the cash flow graph [5]:

- **Totals** display each cash flow as a solid-colored bar.
- **By component** displays each cash flow as a stacked bar, with a different color representing each of the components in the system. Note that penalties and system-fixed costs appear in the graph as "other" costs.
- **By cost type** shows each cash flow as a stacked bar, with each color representing one of five cost types: capital, replacement, salvage, O&M, and fuel. Note that the salvage value appears as a positive value at the end of the project lifetime. For smart grid, grid sales are included in the O&M cost type.

Least cost solar energy is a reasonable figure of merit for smart grid in which renewable energy is the only energy resource. The smart grid yielding least cost can be defined as that showing minimum owning and operating cost over the life of the grid, considering solar energy only. However, the optimum design of a combined solar plus auxiliary energy system based on minimum total cost of delivering energy will generally be different from that based on least cost solar

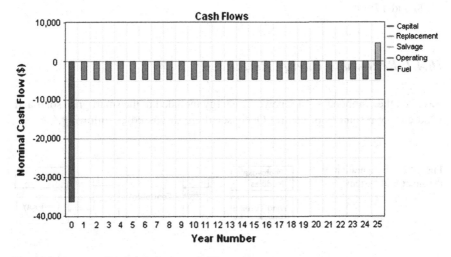

Fig. 10.1 An example of cash flow graph [4]

energy, and the use of least cost solar energy as a criterion is not recommended for the smart grid using solar in combination with other energy sources.

10.3 Design Variables of Smart Grid

The aim of economic design of smart grid is to find the lowest cost system. In general, the problem is a multivariable one, with all of the components in the system and the system configuration having some effect on the performance and thus on cost. For smart grid, the size of components is the most important variable for cost analysis. Therefore, the optimum design needs to find the optimized components' size while the system can deal with the electrical load.

10.4 A Case Study in Central Queensland

The case is of a smart grid consists of both wind turbine and solar PV panel as primary renewable energy component working in parallel with the grid as a backup electricity supplying system. Figure 10.2 shows a general scheme of the system. Figure 10.3 illustrates the proposed schemes as implemented in HOMER simulation tool. It reflects all the components as described in Fig. 10.2.

The load connected to the grid is average 85 kWh/m^2 per day, as shown in Fig. 10.4. It is seen that there is peak of monthly mean load in December and January. This is because air conditioner which is a main electricity consumer is used more frequently in summer, and therefore, there is more electricity consumption in summer. As a daily profile (Fig. 10.5), the peak happens between 13:00 and 17:00 h.

10.4.1 Renewable Energy Resources

Based on the components of renewable energy included in the smart grid, solar and wind energy resources of central Queensland are discussed as following.

Fig. 10.2 Configuration of the smart grid system

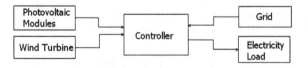

Fig. 10.3 HOMER code of the smart grid system

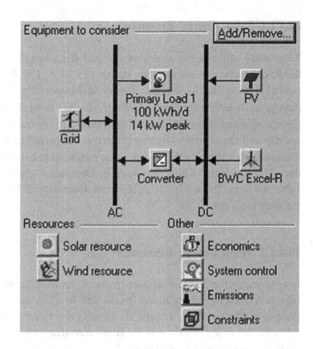

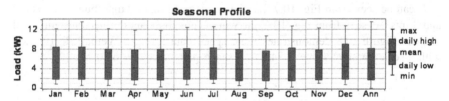

Fig. 10.4 Monthly average load variations

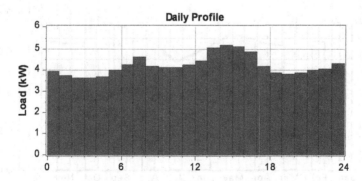

Fig. 10.5 Hourly load variations

10.4.1.1 Solar Energy Resource

Total daily global solar exposure derived from satellite data (MJ/m^2 per day) of the Rockhampton Aero weather station (23.37°S, 150.48°E) for the year 2011 was collected from the Australian Bureau of Meteorology. Scaling was done on this data to consider the long-term average annual resource (5.68 kWh/m^2 per day) for Central Queensland [5]. Figure 10.6 demonstrates the daily radiation in kWh/m^2 per day and the clearness index curve over the period of the whole year. This figure illustrates that the monthly mean solar radiation is between 3.5 and 10 MJ/m^2 per day. The maximum value was seen in December while the minimum one was in July. As seen from Fig. 10.6, Central Queensland has more solar resource in summer than winter [6].

10.4.1.2 Wind Energy Resources

Hourly mean wind speed dataset (m/s) of the Rockhampton Aero weather station which is 10.4 m height above mean sea level is collected from the Australian Bureau of Meteorology. Figure 10.7 shows the monthly mean wind speed between 4.5 and 7.5 m/s [5]. Similar to solar resource, wind resource also show more affluence in summer than winter [5].

It can be seen from Fig. 10.7 that the wind speed mainly distributes between 4 and 8 m/s. As seen from Fig. 10.6, the wind is not abundant during June, July, and August. Figure 10.8 demonstrates wind speed data fitting a Weibull distribution with a scale parameter $k = 1.74$ and a shape parameter $c = 6.30$ m/s [7].

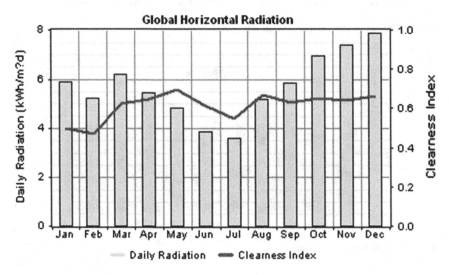

Fig. 10.6 Monthly solar irradiation of Central Queensland

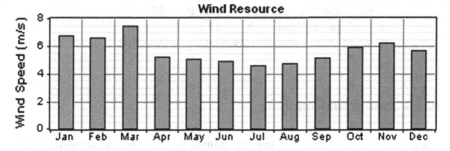

Fig. 10.7 Wind speed for Rockhampton Aero weather station

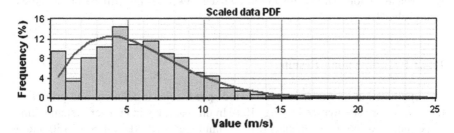

Fig. 10.8 Wind speed probability distribution function for Rockhampton Aero

10.4.2 System Optimization Problem

As the system design was posed as an optimization problem, the objective functions were then formulated corresponding to smart grid constraints and performances [5]. The smart component options are the various configurations combining with different sizes of PV arrays, batteries, and converters.

When producing a specific electricity generation, the optimized system has the lowest costs in the life cycle; therefore, the objective of optimization can be given by Eq. (10.6) [1].

Minimize:
$$f\left(\vec{x}\right) = \left[\left(\sum_{i=1}^{n} kC_i N_i\right) + \left(\sum_{y=1}^{Y}\sum_{i=1}^{n}(O\& M)_i N_i\right)\right] + \left(\sum_{y=1}^{Y}\sum_{i=1}^{n} R_i N_i\right)$$
$$+ \left(\sum_{y=1}^{Y}\sum_{h=1}^{H} E_{g,p} P_{g,p} - \sum_{y=1}^{Y}\sum_{h=1}^{H} E_{g,s} P_{g,s}\right) \tag{10.6}$$

where

C_i	Investment cost (\$)
N_i	Number of each system components (\$)
$(O\&M)_i$	Operation and maintenance cost (\$)
$E_{g,p}$	Electricity purchased from the grid (kWh)

$P_{g,p}$ Price of electricity purchased from grid ($/kWh)
$E_{g,s}$ Electricity sold back to the grid (kWh)
$P_{g,s}$ Price of electricity sold back to the grid ($/kWh)

10.4.3 Results and Analysis

The results of the calculation show the optimal systems and the sensitivity analysis. Considering the electricity price fixed at 0.3 $ /kWh [8], the PV-wind-based smart grid system can be varied to identify an optimal system type for Central Queensland region. The optimization and sensitivity results will be presented as well (Table 10.1).

10.4.3.1 Optimization Results

The optimization results for specific wind speed, solar irradiation, and grid electricity price are summarized in Fig. 10.8. In this case, a wind power system seems to be most feasible economically with a minimum total NPC of $114,130 and a minimum COE of 0.255 $/kWh. This is due to the abundant wind energy resource

Table 10.1 Technical data and study assumption of photovoltaic, wind turbine, grid, battery, and converter [5]

Description	Value/Information
PV	
Capital cost	$4,000/kW
Life time	20 years
Operation and maintenance cost	$3,200/kW
Size options	0, 1, 2, 3, 4, 5, 6, 7 kW
Wind turbine	
Model of wind turbine	BWC Excel-R
Hub height	100 m
Capital cost	$27,500
Lifetime	30 years
Operation and maintenance cost	$25,000
Size options	0, 1, 2
Grid	
Electricity price	0.3 $/kWh
Converter	
Capital cost	$800/kW
Lifetime	25 years
Operation and maintenance cost	$750/kW
Size options	0, 2, 4, 8, 12, 16 kW

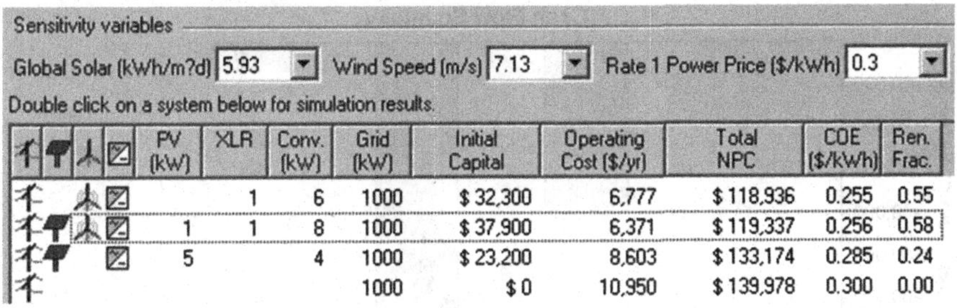

Sensitivity variables

Global Solar (kWh/m?d) 5.93 ▼ Wind Speed (m/s) 7.13 ▼ Rate 1 Power Price ($/kWh) 0.3 ▼

Double click on a system below for simulation results.

	PV (kW)	XLR	Conv. (kW)	Grid (kW)	Initial Capital	Operating Cost ($/yr)	Total NPC	COE ($/kWh)	Ren. Frac.
		1	6	1000	$ 32,300	6,777	$ 118,936	0.255	0.55
	1	1	8	1000	$ 37,900	6,371	$ 119,337	0.256	0.58
	5		4	1000	$ 23,200	8,603	$ 133,174	0.285	0.24
				1000	$ 0	10,950	$ 139,978	0.300	0.00

Fig. 10.9 The optimization results of the simulation

Table 10.2 Cost comparison between standard grid and PV-wind-grid system

Types of Costs	Standard grid	Smart grid
NPC ($)	139,978	119,337
COE ($/kWh)	0.3	0.256

in Central Queensland. In addition, the COE of wind turbine generator is higher than solar array modules.

Figure 10.9 shows that when renewable fraction is 58 %, its NPC is $119,337 and COE is 0.256 $/kWh. This demonstrates that the economic performance of a PV-wind smart grid system is quite similar to the wind smart grid system discussed above. The reduced NPC and COE are just equal to 85.3 % of a standard grid power system. This case has greater renewable fraction (57.6 %) meaning the bigger proportion of renewable energy electricity generations.

The costs of the standard grid and the optimized PV-wind smart grid are compared in Table 10.2. The smart grid system is more economical, while the NPC and COE of the standard grid system are $ 139,978 and 0.3 $/kWh.

Figure 10.10 shows that the cash flow summary of each components of the smart grid. The most costs are used for the wind turbine. In the optimized PV-wind smart grid, the grid component costs the most money ($69,749), while the PV smart gird system just costs $3,876. This is caused by the wind-energy component whose economic performance is better, but solar and the grid component does not. In the PV-wind smart grid, the wind component and the converter cost $39,313 and $6,400, respectively.

The monthly energy yield of each component of the PV-wind smart grid is shown in Fig. 10.11. Implementing under the specific electricity load (85 kWh per day), the PV array produces 1,811 kWh per year which accounts for 4 % of the total electricity generation ($f_{PV} = 4$ %). The wind turbine component produces almost 54 % (24,275 kWh per year) of the system's total energy production (45,306 kWh per year). In another word, the wind generation fraction f_{WT} of this system is 54 %. In this system, the grid purchases account for 42 % (19,220 kWh per year) of the total energy production.

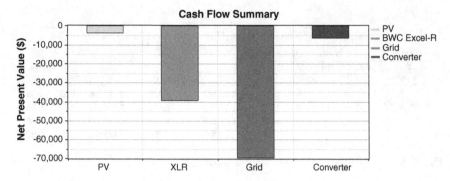

Fig. 10.10 Cost summary of the PV-wind based smart grid

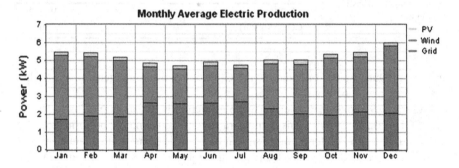

Fig. 10.11 Monthly average electric production of the PV-wind smart grid

10.4.3.2 Sensitivity Results

In this study, sensitivity analysis was done to study the effects of variation in the solar irradiation and wind speed. The long-term implementation of the smart grid based on their respective search size for the predefined sensitivity values of the components is simulated. The emissions, renewable fraction, NPC, and COE are simulated based on the three sensitivity variables: wind speed (m/s), solar irradiation (kWh/m^2 per day), and grid electricity price ($/kWh). For all of the sensitivity values, all the systems in their respective predefined search space are simulated. A long-term simulation for every possible system combination and configuration was done for one-year period.

In this case, solar irradiation is set as sensitivity variables: $G = 4.5, 5, 5.5, 5.93,$ $6.5, 7, 8, 10 \text{ kWh/m}^2$ per day, while wind speed are: $v = 6, 6.5, 7, 7.13, 7.5, 8 \text{ m/}$ s. Moreover, the grid electricity price is also defined as a sensitivity variable ($P = 0.15, 0.2, 0.3, 0.4 \text{ \$/kWh}$). A total of 192 sensitivity cases were tried for each configuration of smart grid. The simulation time was 17 min and 11 s on personal computer with Intel CORE Duo Processor of 2.97 GHz and 4 GB RAM.

The sensitivity results in terms of solar irradiation, wind speed, and grid electricity price analyze the feasibility of each system. Here the feasibility of hybrid renewable energy smart grid system is analyzed based on cost saving. This type of sensitivity analysis of the systems provides information that a particular system would be optimal at certain sensitivity variables. The PV-wind grid system is feasible when the grid electricity price more than 0.3 $/kWh. Under this condition, the renewable fraction can be between 0.55 and 0.65. A PV-wind smart grid system is feasible when global solar irradiation is more than 5 kWh/m^2 per day, and the grid electricity price is more expansive than 0.3 $/kWh. Its renewable fraction can reach between 0.59 and 0.63. When the grid electricity price is constant at 0.3 $/kWh, the PV-wind system is feasible when the global solar irradiation is more than 5 kWh/m^2 per day, and the wind speed is between 6.0–6.75 m/s. Its renewable fraction varies around 0.55.

10.5 Summary

In this chapter, kinds of initial capital cost, replacement cost, and operating and maintenance (O&M) cost are outlined. The economic indicators as optimal objectives and design variables have been discussed.

A case of a PV-wind renewable energy-based smart grid in Central Queensland, Australia has been presented to show the ways of economic analysis and costs savings of a smart grid. A sensitivity analysis is also presented greatly to support the design of a smart grid to meet the demand of costs.

References

1. Liu G, Rasul MG, Amanullah MTO, Khan MMK (2012) Techno-economic simulation and optimization of residential grid-connected PV system for the Queensland climate. Renew Energy 45:146–155
2. Liu G, Baniyounes A, Rasul MG, Amanullah MTO, Khan MMK (2012) Fuzzy Logic based environmental indicator for sustainability assessment of renewable energy system using life cycle assessment. Procedia Eng 49:35–41
3. Duffie JA, Beckman WA (2006) Solar engineering of thermal processes, 3rd edn. Wily, USA
4. Lambert T, Gilman P, Lilienthal P (2006) Micropower system modeling with HOMER. In: Farret FA, Simões MG (eds) Integration of alternative sources of energy. Wiley-IEEE Press, Canada
5. Liu G, Rasul MG, Amannuallah MTO, Khan MMK (2010) Emissions calculation for a grid-assisted hybrid renewable energy system in central Queensland region. In: 4th Network Conference Fusion Solutions: challenges and innovations 2010, Townsville, QLD, Australia
6. Rauschenbach HS (1980) Solar cell array design handbook. Van Nostrand Reinhold Company, New York
7. Burton T, Sharpe D, Jenkins N, Bossanyi E (2001) Wind energy handbook. Wiley, West Sussex
8. Ergon energy (2011) Electricity price. URL:http://www.ergon.com.au/. Accessed 8 June 2011

Index

A

Advanced metering infrastructure (AMI), 30, 33, 34, 115, 158
Association rule mining, 160
Automated meter reading (AMR), 29

B

Bandwidth, 109, 114, 120, 124, 127, 128, 130, 202, 207
Big data, 153, 154, 158

C

Classification, 40, 143, 160, 163, 164, 193, 195–197
Cloud database, 153, 154, 156
Clustering, 40, 143, 160
CO_2 emissions, 35, 37, 46, 71
Communication protocol, 114, 116, 117, 120, 125, 128, 179
Communication standard, 123, 130
Computational intelligence (CI), 39, 192

D

Data integration, 154, 156, 161
Data mining, 40, 151, 152, 159, 160, 161–164
Database, 151–155
Database management system, 152
Dataset, 135, 139, 160, 191, 222
Demand management, 38, 54–56, 135–138, 140, 142, 146, 147, 170, 173
Demand response, 27, 37, 59, 111, 128, 148, 159, 162, 173, 204

Demand scheduling, 135
Distributed generations, 25, 28, 33, 39, 46, 58, 111, 129, 136, 163

E

Emissions, 34–37, 46–49, 71, 137, 226

F

File system (FS), 152, 153
Forecasting, 54, 55, 135, 137–141, 143, 144, 146, 161, 162, 164
Frequent pattern mining, 160

G

Green house gas (GHG), 24, 36, 37, 43, 77, 95, 102

H

Hadoop, 153, 154
Hybrid prediction method, 23, 39

M

Machine learning, 39, 135, 142, 146, 147, 160, 164, 192, 196
MapReduce, 153, 154
Multi-dimensional technical devices, 109

N

NoSQL, 153, 154

A. B. M. S. Ali (ed.), *Smart Grids*, Green Energy and Technology,
DOI: 10.1007/978-1-4471-5210-1, © Springer-Verlag London 2013

O

Outlier detection, 160

R

Regression, 39, 40, 138, 160
Relational database management system, 153, 154, 159, 160

S

Security, 25, 27, 38, 59, 117, 164, 169, 178, 180
Smart grid, 23–30, 35, 37, 39, 43, 57, 58, 174, 196, 197, 202, 205, 217, 220
Smart meters, 25, 109–111, 114
Support vector machine, 40, 135, 138, 160, 164

Printed in the United States
By Bookmasters